Mapping the Code

The Human Genome Project and the Choices of Modern Science

Joel Davis

Wiley Science Editions

John Wiley & Sons, Inc.

New York Chichester Brisbane Toronto Singapore

For my nephews
Jeffrey Wights and Matthew Davis
who will live to see the unfolding results
of the Genome Project
and to understand in a new way
the ancient admonition of the Delphic Oracle:
"Know Thyself."

Library of Congress Cataloging-in-Publication Data

Davis, Joel, 1948–
   Mapping the code : the human genome project and the choices of modern science / Joel Davis.
      p.   cm. — (Wiley science editions)
   Includes bibliographical references (p.   ) and index.
   ISBN 0-471-50383-5.
   1. Human Genome Project.   2. Human genetics—Moral and ethical aspects.   3. Human genetics—Political aspects.   I. Title.
   II. Series.
   QH445.2.D38   1990
   573.2'12—dc20   90-12572

Printed in the United States of America
90 91 10 9 8 7 6 5 4 3 2 1

Preface: Codes and Choices

When I was a small child, I came across a book in the local library about codes and ciphers. I no longer recall the title, but I do remember the feeling that came over me as I read it. It was a feeling of awe, of knowing that there were realities that existed that were not immediately apparent to me. Looking back, it seems that book opened up a whole new world: the world of the hidden, of the unseen, of secret, special knowledge that lay behind a locked door. I knew if I had the key to that door, I could unlock it and find the secret. What's more, I proved it to myself by using the instructions in the book to decode sample messages!

I became fascinated with codes. I constructed my own, and coded different messages that I then passed on to my brothers and friends, to see if they, too, could break the code and read my messages. My best friend Gary King, another towheaded, science fiction-crazy kid who lived around the corner from our house, played with codes for hours at a time—when we weren't reading the latest *Flash* or *Superman* comic books, or trying to construct ray guns from Gary's seemingly endless supply of broken toys and other junk. It was great fun then, and that slightly scary feeling of awe still persists in me today.

You've never attempted to break a code? Try this example:

Uijt jt b tjnqmf tvctujuvujpo dpef.

Get it? The code is a simple one; the rule is: let each letter in the code represent the letter that preceeds it in the alphabet. So "U" represents "T", "i" stands for "h", and so on. The encoded sentence, when decoded using the rule of the code, says:

This is a simple substitution code.

Another way to create a code is to substitute one set of symbols for another. Suppose each letter of the alphabet is represented by a number. A=1, B=2, C=3, and so on to Z= 26. It is then a fairly easy process to encode my name as

10 15 5 12 4 1 22 9 19

There are many different ways of coding information. One of my favorites as a child was a rune code, in which different Celtic-type runes stood for different letters. A box might represent the letter "c", while a box with a dot in the middle stood for "q", and something like this ⌐ stood for "z". A message written in rune-code looked absolutely mysterious and strange, like nothing normal at all. I loved it.

However, most of us left that enthusiasm behind as we grew older, along with our interests in *Flash* comic books, ray guns, live frogs, and monsters hiding in the attic. A few people did not; they grew up to become experts working for the CIA, or computer programmers, or perhaps geneticists.

For the past four decades, molecular biologists and geneticists have been unraveling one of the greatest and most basic of all codes: the genetic code. And it is a marvelous code indeed. With only four "letters," this code spells out all the instructions to create every form of biological life that exists, has existed and in all likelihood ever will exist on Earth. From the simplest cyanobacteria to the largest dinosaur, from ferns to worms, from hawks to humans, the genetic code spells it all out. What's more, there is enough flexibility in the code to allow for nearly infinite variations within each life-form, and for the relentless pressure of environmental change to create totally new species of life out of older ones.

The totality of the genetic code for a particular species is called a *genome*. In the last few years, enough information and technology has accumulated for researchers to begin seriously working at decoding the entire genomes of different species. The ambitious goal of this work, the Genome Project, is to read the entire genetic "encyclopedia" of a species, in particular of the one called *Homo sapiens*—humans.

It is not inaccurate to say that every language is a kind of code, and the genetic code is a form of language, the language of life. It is a series of words and sentences and paragraphs which, when spoken aloud, make a living creature.

A map, too, is a kind of code: a set of symbols and a set of rules for interpreting the symbols, so that they can carry information. The information in this case is in the form of a picture. A map of Olympia, Washington, where I live, is not the same as the city itself. But it does represent, more or less accurately, that city and its streets, buildings, and landforms. The genetic code is also a kind of map. At least, it can be represented in the form of maps. Such a map might be a pictorial description of the information contained within the code, or of the physical location of different "letters" and "phrases" within a gene or chromosome.

When Gary King and I played with our codes as children, we never gave a thought to the rightness or wrongness of what we were doing. We may have written delightfully obscene words in code, but that was probably the extent of our journey into the realm of codes and sin.

In point of fact, a code has no intrinsic moral or ethical value. There may be some such value to the message it contains, either absolutely or relatively. Or, the information contained within the code might be used for good or evil purposes. The code itself, however, is value-neutral.

The same is true of the genetic code, and of the effort to map it out. The genetic code is an incredibly powerful one, to be sure, but with no moral or ethical value to it. It simply encodes information. The process of mapping the code will present humanity with an enormous amount of information, and a new depth of knowledge about the way living creatures work.

The cliché that knowledge is power is usually true. It is also true that power can be used for good or for ill. How the power gained through the mapping of the human genetic code is used will depend not just on the researchers and bureaucrats, but on you and me.

The human genome is *our* genome—mine, and yours. It belongs to no one. And to everyone. It is a creation of 4.5 billion years of evolution. The Genome Project could be the ultimate violation of privacy, or it could be an extraordinary doorway to renewed life, to health, to healing.

Mapping the Code is the story of the Genome Project, of how it came to be, of who the people were—and are—who are engaged in this effort to map the human genetic code, of the ethical and moral questions and choices which are already being asked as a consequence.

May this book help you, in at least some small way, to understand it. To ask questions. To make your own choices.

Acknowledgments

It would not have been possible to write this book without the help of many people. My thanks to those involved in the Genome Project with whom I was able to talk, including George Bell, Paul Berg, Christian Burks, George Cahill, Jr., Anthony Carrano, Robert Cook-Deegan, Charles DeLisi, Russell Doolittle, Diane Hinton, Robert Jones, David Kristofferson, Suzanna Lewis, Tom Marr, Victor McKusick, Benedictus Nieuwenhuis, G. Christian Overton, Peter Pearson, Maxine Singer, Robert Sinsheimer, Clay Stephens, David Smith, Edward Thiel, Anne Marie Wan, and Raymond White. Their gift of time and knowledge was most helpful. Any errors herein are not theirs, but mine.

I also wish to thank my agent, Joshua Bilmes, and my editor, David Sobel, for their support and assistance on this project. The librarians and staff of the Washington State Library, University of Washington Health Sciences Library, The Evergreen State College Library, and the Olympia branch of the Timberland Regional Library were (and are always) of inestimable assistance.

For help and assistance of various other kinds, I wish to thank Mark and Nancy Allen, Dick Batdorf (Fare forward, Voyager!), Astrid and Greg Bear, Rebecca Carter, Gerald and Toni Davis, Grant Fjermedal, Tom Foote, Elana Freeland, Carla and Dean Jones, Carol Knobel, Vicki Kreimeyer, Mary Giles Mailhot, OSB, the Nisqually Dance Circle, Fred and Megan Ogden, Barbara Park, Annie Scarborough, Joan Weeks, and Brooke Wickham.

And to She who moves at the center: thanks again.

Special Acknowledgment

Portions of Chapters 5 through 8 are heavily based on the authoritative reporting of Leslie Roberts, senior writer at Science *magazine, in a series of articles on the genome project which appeared in* Science *from July 31, 1987 through January 19, 1990. The author acknowledges his debt to her work.*

Contents

1

~~~~~~

The Territory, and the Map

Journey through all the universe in a map,
without the expense and fatigue of travelling,
without suffering the inconveniences
of heat, cold, hunger and thirst.

MIGUEL DE CERVANTES
Don Quixote de la Mancha

WILLAM SHAKESPEARE said it first, through the character Miranda in *The Tempest*: "O brave new world, that has such people in't!" Within the context of the play the line is a joyful one. It was Aldous Huxley who appropriated the first part as the title of his classic science fiction novel about a utopian future that is not all it seems.

That is the way of all futures, be they possible or inconceivable. Imagine what the future might be, considering our relentless drive to shape it. Shape it we do—by accident or by deliberate intent.

But the future never quite seems to turn out to be the same future we imagined. Keep that in mind, for we are now embarking on a journey through the past and present to the future—the future of the Genome Project.

AT THIS VERY MOMENT, researchers in laboratories around the world are compiling the first detailed maps of the biomolecular essence of a human: the genetic code. Their ultimate goal—a complete map of the entire human genetic code—may be reached in the first decade of the twenty-first century. In the last several years, this effort has been formalized in a series of scientific and government meetings and symposia. It goes by the general name of the *Genome Project.*

A *genome* is defined as all the genetic information of a particular species. We can speak of the human genome, the mouse genome, the dog genome, the housefly genome, and so on. Genetic variations in the genome, various combinations of different possible genes—like different spellings of the same words—create the infinite variety that we see among individual members of a species. With this in mind, however, the genome of the author of this book is still very similar to that of the book's editor—both are members of the species *Homo sapiens.*

The Genome Project (referred to by some as the *Genome Initiative*, or the *Human Genome Initiative*) is not really a single, massive program. It is actually a collection of interrelated, coordinated projects; some have just begun, others have been going on for years in university, commercial, and governmental laboratories. Many are being coordinated and overseen (loosely) by federal agencies of the United States, agencies and ministries of other nations, and at least one international organization.

The Genome Project is already revolutionizing the way biologists do biology. It is leading traditional scientists in messy laboratories and intense small research groups into the world of Big Science—and into the world of Big Science budgets and politics.

It will do much more than that.

The Genome Project will revolutionize the practice of medicine. It will utterly change the field of genetics. It may totally reshape the nature of *Homo sapiens*.

And it will force everyone, scientists, laypeople, priests, and parents, to grapple with perplexing moral and ethical questions: Does the right to privacy extend to one's genetic code? Is that right absolute? If not, under what conditions can someone else—for example, one's government, doctor, insurance company, or employer—have access to one's personal version of the human genome? Once we understand the genetic basis of everything from high blood pressure to schizophrenia, from Huntington's disease to musical genius, are there conditions under which it would be ethical to alter the genes in our ova or sperm and thus change the shape and characteristics of future generations?

Even today as the Genome Project is just getting underway, some medical practices created by the genetic engineering revolution are coming under increased scrutiny and debate. It is now possible to determine the presence of genetic markers for some genetic diseases in a fetus. But is it ethical for a woman to abort a fetus afflicted with a genetic disease such as cystic fibrosis, which always promises painful death in childhood? If the fetus has the genetic marker for Down's syndrome? If it has genetic markers for a high tendency towards diabetes? Where do parents draw the line? Genetic counseling clinics, once the figment of a science-fiction writer's imagination, are now being set up in hospitals and medical centers across the country. Does a couple have the right to know before conception what the genetic makeup of their child might be? How will such tests be funded? If it's pay-as-you-go, wouldn't the poor be excluded from the benefits of genetic counseling? Is not this a form of *eugenics*, the practice of genetically tailoring a population to fit a certain set of criteria?

What exactly will the future be, this future that unfolds from the Genome Project in the same way the instructions for making a human unfold from double-helix DNA? No one can say for sure. Here, though, are three possible futures. Imagine them as three tiny threads pulled from the huge tapestry of History That Might Be that is woven with the warp and woof of the Now in which we live. These futures are not far away; they may come to pass during lifetimes of most who are reading these words.

⊃⊂ FUTURELINE 1: AT THE COUNSELING CENTER

Harris and Callie Webster waited in the sitting room of the Lacey Genetics Counseling Center. They paged through old copies of Readers Digest *and* Hippocrates, *looking at but not really seeing articles like "Ten Ways to Save a Failing Marriage" and "Cystic Fibrosis and the Nobel Scandal." A month earlier they had begun having their genomes mapped. First there had been an hour's lecture and discussion of how the mapping would be done, what it is designed to show, what the different ethical positions on the procedure are, and what its possible consequences are. Then came the procedure itself. It had been simple: Some blood samples had been taken from their earlobes, some tissue scrapings from their palms. Harris had submitted a semen sample in a test tube supplied earlier. Callie had had a medical procedure in a small but comfortable room. Using local anesthesia, the doctor, using the techniques perfected for in vitro fertilization, had retrieved an ovum.*

Now came the payoff, a complete readout of their genetic codes. In particular, the maps would look for the presence of genetic markers for diseases known to be caused by genetic mutations. Huntington's disease, cystic fibrosis, Down's syndrome, Tay Sachs disease, sickle-cell anemia (the list was a long one, and grew each month). As the doctor had pointed out the month before, most parents have perfectly healthy children. The point was not to be frightened, she said. "The point is to have a better idea of who you are genetically, so you can make better-informed decisions about having a child. The chances are that the news you'll get the next time you're here will be completely good."

"Callie and Harris?"

The two looked up from their magazines to see the doctor's face peeking around the door. "Come on in," she said.

"First things first," the doctor said, handing each of them a computer disk and a large manila envelope. "That's your genetic map, in hard copy and on disk. Frankly, you'll probably find most of it incomprehensible, unless you're geneticists. But there's a part at the beginning that summarizes the results in plain English."

"And they are?" asked Callie.

"Well," she said, "it seems you, Callie, carry the gene marker for Duchenne's muscular dystrophy or DMD."

"Does it mean we can't have children?" Harris asked.

The doctor shook her head. "No. It doesn't mean you can't have children. You can. Well, look, DMD is a muscular disease that occurs in one in every 3,500 or so male infants. The gene that causes it is on the X chromosome. Remember from our talk last month that the X chromosome is one of the two sex chromosomes. The other is the Y chromosome."

"Right," said Callie. "If you get two Xs, you're female, and if you get an X and a Y, you're male. And you can't get two Ys, because the woman always contributes an X, while the man contributes either an X or a Y."

The doctor smiled. "You've got it," she said. "The gene for DMD is located on the X chromosome. Its location is named Xp21.2. It's very rare for a female to come down with DMD, because both of her X chromosomes would have to have the gene. For boys, it's different. If they inherit an affected X chromosome, they get DMD. It won't happen every time, of course. Harris, your X chromosome is normal. It's Callie who has an X with the DMD gene."

"What are the chances that a fertilized egg will inherit it?" Callie asked.

"Well, there are four possible combinations. A normal child would result from Harris's X and your normal X, or from Harris's Y and your normal X. Harris's X and your X with the DMD gene would result in a normal female infant, because the normal gene on the X chromosome from Harris will produce the proper proteins, and compensate for the deficiencies of the DMD gene on the X chromosome from you. The problem is a combination of Harris's Y chromosome with your DMD gene-carrying X. There would be no normal X chromsomome to compensate for the DMD chromosome. The result is a male child—it's X and Y, remember—with muscular dystrophy."

Harris asked, "Is it fatal?"

The doctor nodded, her smile gone. "Yes. It still is. It would begin in early childhood, about when he would be learning to walk, and then the muscle wasting progresses. He would most likely be confined to a wheelchair by age 12. It is very rare for someone with DMD to survive past the age of 20.

There's another genetic disease called Becker's muscular dystrophy, which is much milder than Duchenne's. It's caused by the same gene, but the gene changes that cause Becker's are different from those that cause Duchenne's. I truly wish that the marker we found was for Becker's, Callie. Actually, I wish it weren't even there at all."

"What do we do now?" Callie asked.

"If Callie gets pregnant," said Harris, "there's a one-in-four chance that the baby will have the gene for this. Right?" He shook his head angrily and spat, "Jesus! What kind of odds are those, anyway!"

"We could have it aborted," replied Callie. "We've talked about it before. It's an option."

"An option," he said flatly, "an option."

"It's not just an intellectual thing anymore, is it." She began to cry, silently.

The doctor walked to the front of her desk and took their hands in hers. *"Look. I can't tell you what to do. I can give you the medical facts; I can even give you my own personal opinion, if you want it; and if you wish, you can talk with Dr. Rice, the ethicist on our staff. He can give you a lot of help in making your decision.*

"But the decision is yours."

Harris sighed deeply. *"You know, I work with computers,"* he said. *"Have for years now. I've always told people how important they are, how they've made this the Information Age. The more information the better, I've said. The more information you have, the better decisions you can make. "But now, it seems that sometimes, having more information just means that your decisions are . . . well . . . deeper."*

⊃⊂ FUTURELINE 2: THE GENEBUM

It is night in the city nestled by Elliot Bay. From the window of a passenger jet, Seattle looks like the crown jewels, with its proud towers alit and glowing.

Down on the streets near the waterfront, things don't look so pretty. Garbage chokes many of the alleys. The city that gave the world the phrase "skid row" has one of the largest in North America, and it continues to grow. The homeless, the alcoholics, the drug addicts, the prostitutes have been joined by the genebums.

Who are the genebums? They're people like you. You are nineteen, female, with brown hair, blue eyes, dark skin. Very thin. You sometimes eat your meals from dumpsters. The only jobs you can get are gruntwork: Bussing tables in seedy restaurants, picking fruit in the Eastern Washington orchards or vineyards, backbreaking flood-control duty on the Olympic peninsula in winter, chopping firebreaks in the Okanagan in the summer.

The problem is not that you can't work at good jobs. You can. You would love to work. You're extremely intelligent, and have a high creativity quotient as well. But no one will hire you, at least not for *any job with a future. One of the first things that prospective employers do is plug your name into the National DataBank, pull up your GenRecord, type in a few commands, and on the screen flashes:*

"OMIM #17389: 19q13-ter. Susceptibility to polio."

Your GenRecord, like everyone else's, contains a readout of your entire genetic code—all 123,547 genes that make up the human genetic repository, the human genome. By law, the amendment to the Social and National Security Act of 2001, each person's individual genome is mapped and stored in one of eight regional computer databases. By law, the information is publicly available to everyone—including you, your doctor, your friends and enemies, your potential employers, your potential insurance company.

Your genome has betrayed you. On the "q" arm of your nineteenth chromosome is a gene that makes you susceptible to contracting polio. It doesn't matter that everyone gets vaccinated for polio and has for nearly a century. The virus is alive and well, and is usually found in the sewers and effluent of every city, town, and hamlet on aching old Terra. Sometimes, the vaccine doesn't work. What employer is going to take the chance that you might be one of those people for whom the vaccine is useless? If you were to contract polio, your employer or insurance company would have to pay your medical costs.

Of course, no one has a perfect genome. At least, that's what the vids say. You wonder about that, though. There are lots of people living in those brightly lit towers. They have lots of money. And Seattle is home to the Genomics Corporation (GC), one of the megagiant biotech companies. You suspect that more than a few of those folks have slipped a few large bills under the table to GC, and had some gene surgery done. Not just to their own somatics, either, but to their wigglers or eggs, too—Why not? If they can afford the very best, why not do it for their descendents, too? So it's illegal? Those with money make the rules.

Oh, you've tried a few times to hook onto some job that pays more than a few hundred dollars a week and doesn't break your back. You've still got some decent clothes stashed away in the broom closet of a room you rent on Second Avenue. But you've pretty much given up.

Tonight, you are tired. And angry. And desperate. And at the end of your rope.

Tonight, you hide in the shadows near the Westlake Towers Hotel, waiting for some rich, unescorted young businesswoman to walk by.

Because tonight, you are going to take revenge.

⊃⊂ FUTURELINE 3: THE CHOICE

Some things are always the same. Like the waiting rooms in doctors' offices. David Murrill sat in one with his wife Miriam. They wanted a child. Two weeks ago, David had donated sperm, Miriam some of her ova. However, since then they had gone through genome-mapping of their germ cells. This was the moment of truth.

Their doctor, a prim but pleasant looking woman walked into the room. "Are you the Murrill family?" she asked. "That's us," replied Miriam. The doctor smiled. "I'm Eileen Garrison," she said, coming over and shaking their hands. She gestured with her head toward a door. "Come on into my office, and I'll give you our Official Genome Chat and Peptalk."

The tension in the room broke, and they laughed. In the office, however, the doctor looked a little more somber.

"Actually," she said, "this isn't quite the standard chat and peptalk I have for you. But it isn't quite bad news, either."

David frowned. "I don't understand. Is there a problem?"

"Yes, and no. We've mapped and sequenced the DNA sequences for your sperm and your eggs," the doctor said, looking first at David, then at Miriam.

"For the most part, you have good, healthy genomes. The one problem you do have is very serious, though. You both carry the gene for Huntington's disease on chromosome 4."

"Oh no!" cried Miriam, as David took her in his arms.

"OK, hold on, now. Hold on," Dr. Garrison continued. "It's not as bad as you think. We can do something about it."

David looked up, his face still pale. "What?"

"Genetic surgery. We call it 'cutting and pasting.' It works, but it's not cheap, at least not now. Five years from now—maybe three,

if some of my friends are right—it'll probably be covered by your universal health insurance. It's a painless procedure that involves re-placement of the gene that causes Huntington's with a nearly identical one that does not."

Miriam asked, "What do we have to do?"

The doctor smiled, "All you have to do is drink some orange juice and call me in the morning. Really!

"It's not surgery in the conventional sense. The cutting and pasting is done with two specially constructed viruses. One virus is engineered to zero in on the Huntington's gene and snip it out of chromosome 4 with an enzyme, which acts as a chemical scissors. Then the virus temporarily plugs the hole with another molecule. The second virus is 'infected' "—she drew quotation marks in the air—"with the replace-ment gene. This virus then homes in on the plug in the chromosome, carries the new gene to it, and inserts the gene into exactly the right position. Voilà! Cut and paste. Huntington's gene cut. New gene pasted in. The virus also passes through breast milk to your baby, and the same process takes place.

"If you agree to the procedure, I will order sixteen vials of virus, and you can pick them up at the pharmacy here in a week. The viruses are contained in a suspension of ordinary distilled water. You will each get two sets—four vials of the first virus and four of the second. Mix the contents with a glass of orange juice and take one glass each morning for eight days—four for the first set of vials, four for the second.

"The reason for so many vials of virus is, frankly, overkill. The contents of two vials should do the trick. But we want to insure infection of every cell by both viruses, and we want to insure replace-ment of every defective gene in your infant with a normal gene. And one other thing—this treatment not only guarantees that this child will not contract Huntington's disease, but also that the next one won't, either."

"What about side effects?" Miriam asked.

"Good question. You'd be surprised how few people ask. The known side effects are diarrhea from about day 3 through 7, and

rarely, 1 in 100,000 contracts a vaccinia infection. That's because vaccinia is the carrier virus. Vaccinia infection is not fatal."

"Is there any other way to treat this?" asked Miriam.

Dr. Garrison looked puzzled. "No. This is the only way."

"I'm sorry, then," said David. "We must refuse. Our religion forbids this kind of genetic medicine. Our Teacher and Prophet has clearly stated that this kind of treatment is immoral."

ANALOGIES

One easy way to grasp the nature of the genome is to think of it as the surface of a planet. A planetary surface is made of many different parts—mountains, plains, marshes, lakes, rivers, oceans, continents, and so on. From a great distance, one can see the Earth as a whole but can not make out the details of its surface. From a closer viewpoint—say, a geosynchronous orbit some 22,300 miles out—one can still see the entire planet but can also make out more details. Even closer in—say from a shuttle orbiting a few hundred miles out—an observer can see many surface details, but cannot see the entire sphere. Finally, from an airplane flying just a mile or so above the surface of the Earth, an observer can make out extremely fine details.

Geneticists today are still like astronauts looking at the Earth from a moderately far distance. They can see the planet in its entirety, and can make out some of the details of its surface. But they are still not close enough to see the fine details—rivers, the Grand Canyon, or fields of corn and wheat. They certainly cannot see something as small as Hicks Lake, which lies near the town of Lacey, Washington. And there is no way they can make out city streets or find individual homes in Lacey.

The Genome Project will produce maps of the "DNA planet." These maps will have different levels of detail, from the whole planet to the location of a duplex in Atlanta, Georgia.

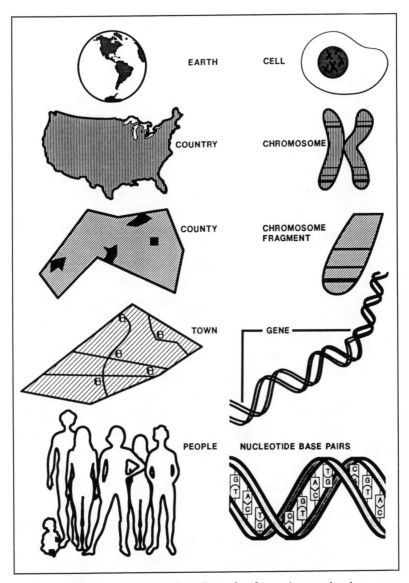

One way of understanding the scale of genetic mapping is to compare it to different maps of the Earth. Chromosomes are analogous to countries, chromosome fragments to states or counties, genes to cities or towns, and base pairs to individuals.—Reprinted from: U.S. Congress, Office of Technology Assessment, "Mapping Our Genes—The Genome Project: How Big, How Fast?" OTA-BA-373 (Washington, D.C.: U.S. Government Printing Office, April 1988.)

Now consider the genome as an encyclopedia that contains all the instructions necessary to create a living person. The human genome contains about a hundred thousand "entries" found in one of 24 different "volumes." These entries are written with an alphabet of only four letters that are repeated some three billion times in about twenty combinations of three-letter "words." All of this information is packed into an area much smaller than the naked eye can see. This is an encyclopedia that can literally fits on the head of a pin, yet it guides the complex development of a human being.

CENTIMORGANS AND GENETIC MAPS

The ultimate purpose of the Genome Project is to identify and locate all the genes in all the chromosomes of the human species. It will do that by creating maps of the genome, a representation of the universe within us, of the 100,000 genes that make humans human. The maps will be both on a plane surface, such as a page of a book, and on an electronic surface within databases and computer terminals.

We measure the distance between two points on Earth in miles or kilometers. On maps those distances are scaled down to miles per inch kilometers per centimer. Distances between genes in the genome, however, are measured differently. One way is simply by counting the base pairs between genes, which is still nearly impossible, since we know the existence and location of merely one half of one percent of the human genes. Another way is by using *units of probability* called *centimorgans* (*cM*), named after Thomas Hunt Morgan, the great geneticist. Each centimorgan represents a 1 percent chance that two genes will separate when chromosomes recombine during *meiosis*, or cell division. It is obtained from the cross-over frequency. For example, suppose a known genetic marker is inherited with the gene for Huntington's disease 96 percent of the time. Thus this marker and the gene become separated (cross over) during meiosis about 4 percent of the time. This means the genetic signpost for Huntington's disease is 4 cM from the gene.

Is there some way to translate probablity units into units of physical length? Yes, partly. The average human chromosome contains about 120 million base pairs. If genetic sites separated by 1 cM have a 1 percent chance of a break occurring between them, then a centimorgan must be 1 percent of 120 million base pairs—roughly 1.2 million base pairs. It must be assumed, however, that the probability of a break depends *only* on the distance between the genes. This isn't true all the time, so the correspondence between numbers of base pairs and the measurement in centimorgans is never quite exact. But it is close enough. It also shows that in a genome 3 billion base pairs long, a gene at a distance of a million base pairs away is pretty far.

There are two major kinds of genetic maps: genetic-linkage maps and physical maps. A *genetic linkage map* shows the location of a gene or a cluster of genes relative to second genetic locus on the basis of how often they are inherited together. A *genetic locus* can be a gene itself or some piece of DNA with no known function but whose inheritance pattern can be determined. The primary unit of measurement for genetic linkage maps is the centimorgan.

A *physical map* represents the physical location of identifiable landmarks on the DNA strand. The major unit of measurement for physical maps can be anything from banding patterns to actual base pair counts.

Within this general format are several specific kinds of genetic maps, some that exist and others that have yet to be developed.

■ *Cytogenetic Map.* Victor McKusick of Johns Hopkins University School of Medicine has been creating and refining cytogenic maps for at least a decade. The cytogenetic map shows the physical locations of banding patterns created with chemical stains much as a political map of the world shows the locations (and shapes) of the fifty states of the United States.

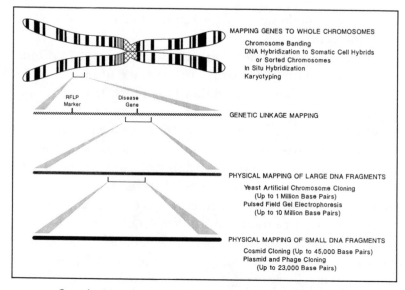

Genetic Mapping at Different Levels of Resolution.—Reprinted from: U.S. Congress, Office of Technology Assessment, "Mapping Our Genes–The Genome Project: How Big, How Fast?" OTA-BA-373 (Washington, D.C.: U.S. Government Printing Office, April 1988.)

Such a political map, however, is of no help in finding 42nd Street in New York City. Its resolution, is very coarse: A cytogenetic map also has a coarse resolution. Chromosome staining creates about ten bands per chromosome; if the average chromosome has 120 million base pairs, each band has about 10 to 12 million base pairs within it. Gross chromosomal abnormalities can thus be located to within 10 million base pairs (1×10^7 bp) with a cytogenetic map.

■ *Restriction Map.* Dr. Raymond White of the Howard Hughes Medical Center at the University of Utah is one of the leaders in creating restriction maps of human chromosomes. This kind of genetic linkage map uses genetic markers or "signposts" called *restriction length fragment polymorphisms (RFLPs)* to show the probability of genetic markers' being close together. RFLPs are fragments of DNA cut from the same place in the same chromosome of different people. The cutting is done with

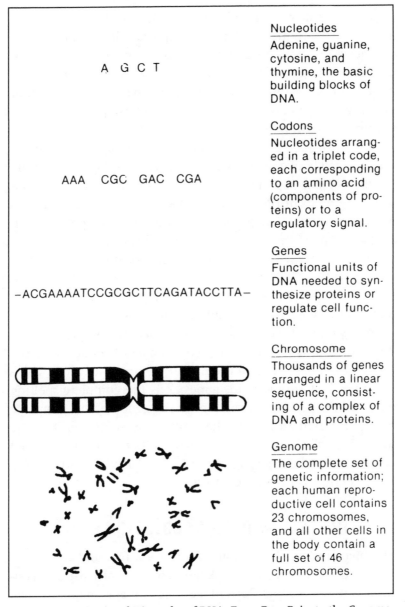

Nucleotides
Adenine, guanine, cytosine, and thymine, the basic building blocks of DNA.

Codons
Nucleotides arranged in a triplet code, each corresponding to an amino acid (components of proteins) or to a regulatory signal.

Genes
Functional units of DNA needed to synthesize proteins or regulate cell function.

Chromosome
Thousands of genes arranged in a linear sequence, consisting of a complex of DNA and proteins.

Genome
The complete set of genetic information; each human reproductive cell contains 23 chromosomes, and all other cells in the body contain a full set of 46 chromosomes.

Organizational Hierarchy of DNA, From Base Pairs to the Genome. Reprinted from: U.S. Congress, Office of Technology Assessment, "Technologies for Detecting Heritable Mutations in Human Beings" OTA-H-298 (Washington, D.C.: U.S. Government Printing Office, September 1986.)

a chemical called a *restriction enzyme*—hence the name of the map. Restriction maps using RFLPs (pronounced "riflops") incorporate information gathered from the genetic inheritance patterns of large multigenerational families. This kind of genetic map has a resolution about ten times higher than that of cytogenetic maps. One can identify a base pair location to within a million base pairs (1×10^6 bp).

■ *Cosmid Map.* The cosmid map is a physical map that gives the distance between genes and base pairs proportional to the number of bases. It uses information from overlapping *cosmids*—segments of DNA about 40,000 base pairs long. The resolution of a cosmid map is ten to a hundred times greater than a linkage map, and at least a thousand times greater than a cytogenetic map. It can pinpoint the location of genes to within 10 to 100 thousand base pairs of each other (1×10^4 to 1×10^5 bp). Physical maps do not yet exist, but the first ones will probably be available in the early 1990s.

■ *Sequence Map.* McKusick has called the sequence map "the ultimate map." It is the actual physical sequence of the three billion base pairs of the human genome—of all 46 chromosomes and 100,000-plus genes. It is the complete spelling out of all the genetic instructions in the genome. The sequence map of the human genome has not yet been created. It is the ultimate goal of the Genome Project. When it *will* be finished is uncertain. Guesses range from the mid-1990s to some time in the early twenty-first century.

THE GENOME PROJECT OBJECTIVES

The objectives of the Genome Project are five-fold:

1. To create a genetic linkage map with at least one centimorgan resolution. Current genetic linkage maps are five times as coarse as this.

2. To create a physical map of each chromosome within five years. This will include a *clone respository* (that is, a "library" containing copies of the cloned cosmids used to make the map)

and a *reconstruction cookbook* that will enable other researchers to duplicate the physical map of a chromosme in their own laboratories.

3. To improve the technology for sequencing segments of DNA, and for handling and storing DNA.

4. To obtain the actual sequences of large areas of the human genome—that is, the construction of at least partial sequence maps of the human genome.

5. To develop new and effective computer databases for the first, second, and fourth objectives in order to handle the immense influx of data that will result from mapping the human genome.

There is a great deal of variability among the genetic codes of individual humans, as much as one base pair per thousand. Since the human genome contains up to 3 billion base pairs, the genetic differences among individual humans is as much as 3 million base pairs of DNA. So it would be foolish to call the mapping of one person's DNA "the archetypal human genome." The actual genetic maps that will be created will not be of any particular person. No individual will be "immortalized" as "the archetypal human" with "the archetypal genome." Rather, specific chromosomes and parts of chromosomes from many different individuals will be mapped, and the map of a chromosome will actually be a synthesis of many individuals.

The genomes of several other species will also be mapped including bacteria and other creatures that play an important experimental role in genetics:

■ *E. coli*: This bacteria is one of the most important experimental creatures in genetics. A genetic linkage map of the genome of *E. coli* has been partially completed. There is as yet no physical sequence map.

■ *Yeast*: The mapping of the yeast genome is already underway; much of it is already mapped.

■ *Nematode*: The mapping of the nematode genome is underway; but there is still much to be done.

■ *Mouse*: The genome of the mouse, a standard experimental animal in genetics, may be the single most important nonhuman genome to be mapped. Small pieces of it have been mapped simply as a by-product of its application in other genetic research. However, there are years of work ahead before the genome of *mus musculus*, or mouse, is totally mapped.

THE LONG ROAD AHEAD

Mapping the human genome will not be easy, even with advanced computer technology and new automated DNA-sequencing techniques. Though a start has been made, there is still a long way to go. The current status of the Genome Project clearly proves this:

■ A 5 centimorgan genetic linkage map is now available at Yale University.

■ Some attempts to agree on a common computer language and database format are underway. Up till now, there has been no central database of all clones and cosmids needed for a physical map; many investigators have been using their own methodologies and nomenclature.

■ About 30 million DNA sequences are collected in the computerized GenBank, located at the Los Alamos National Laboratory in New Mexico. That is less than 1 percent of the human genome, for only 10 percent of those 30 million DNA sequences are of human DNA. The rest is of different experimental animals and bacteria, such as mice and *E. coli*.

■ Researchers still do not have computational tools powerful enough to deduce the most likely map from the raw data. Biologists and geneticists also need computerized tools to enter the data into a database, to manage the databases and to make them available, and to analyze the data in the databases.

In fact, the entire field of biology and genetics is still woefully undercomputerized, due largely to the very nature of biology. It's a wet and messy science done with pens and notebooks in crowded laboratories, a far cry from the general public's vision of antiseptic, white halls of gleaming machinery. Putting computers into this environment at first seem ridiculous. But this is largely a problem of attitude as many older biologists must grapple with learning to accept and operate computers.

The bottom line for the Genome Project is the absolute necessity of computers. The amount of raw information that will be produced by mapping and sequencing 3 billion base pairs of human DNA will be overwhelming. The only way to store, categorize, and use that data will be to use computers.

In fact, one of the major justifications for the initial involvement of the U.S. Department of Energy in the Genome Project was to encourage the development of supercomputers.

2

In the Beginning

Man can learn nothing unless he proceeds from the known to the unknown.

CLAUDE BERNARD

THE DESIRE TO create a map of the human genetic code—a physical, graphic representation of the genetic information that makes us human—has it roots in our very humanity. Only members of the species *Homo sapiens* are capable of asking "Who am I?" Only we are able to say to our fellow humans, with the Delphic Oracle: "Know thyself." The drive to map the human genome really begins with those two statements.

It was not until the beginning of the twentieth century that we began to acquire the knowledge and the tools with which to answer the question. The first steps to the Genome Project were taken with the rediscovery of the work of a nineteenth-century Austrian monk who became fascinated with peas.

MENDEL AND HIS PEAS

Gregor Johann Mendel (1822–1884) was a member of the Roman Catholic religious Order of St. Augustine. During much of his later life he lived and worked at the Augustinian monastery in Brno in what is now Czechoslovakia. He eventually became the community's abbot. From 1843 to 1868 Mendel carried out a detailed series of experiments on garden peas and other plants in his garden. He used carefully controlled pollination techniques, kept track of the outcome, and carried out a statistical analysis of the results. His years of experimental work produced the first scientific evidence for the process of heredity—the transmission of certain characteristics of a living creature from one generation to the next. He published the results of his work in 1866, but it was almost completely ignored, and fell into obscurity. Disappointed but not crushed, Mendel spent the last years of his life focusing on his vocation as a monk.

Mendel's work literally changed the shape of the world, of his future and ours. That work depended mostly on his observations of different attributes of garden peas, such as height, seed color, a smooth or wrinkled surface, and so on. Mendel deduced that seven of those attributes were controlled by seven pairs of what he referred to as "unit characters."

An example of the kind of experiments Mendel did illustrates the way he discovered the laws of inheritance. Mendel noticed that the color of the seeds included two unit characters, green and yellow. He cross-bred parent plants of the two varieties. The seeds of the resulting hybrid were all yellow. This meant that the unit character for yellow was more dominant than the unit character for green. Mendel then inbred these hybrid plants. The offspring of the hybrids had both yellow and green seeds. There were three yellow seeds for each green one.

This result can be simply explained if each parent plant contributes to an offspring one inheritance unit for seed color. It could be a unit for yellow color (capital C for the dominant *COLOR* yellow) or the unit for green color (small c, for the recessive *color* green). Four possible combinations of two color units can result: CC, Cc, cC, or cc. Three of every four offspring have yellow seeds because they receive at least one C unit of seed color, the unit character for yellow. Yellow is dominant. One out of four offspring have green seeds because they do *not* receive a C unit character. Instead they receive two c units. Only this offspring can have green seeds, since green is a recessive color.

If this theory is correct, then we can make a prediction about what proportion of seed colors will result from different breeding schemes. Suppose, for example, we take two of the hybrid pea plants with green seeds and inbreed them. We can predict that their offspring will always have green seeds, and never yellow seeds. If the plants have only c inheritance units, this will be true. And in fact it is. Suppose, however, we take a hybrid plant with green seeds and one with yellow seeds, and inbreed them. What color seeds will their offspring have? Now it gets more complicated. Each parent has in its germ cells one color inheritance unit from its pair of units. A hybrid *CC* will always contribute a dominant yellow *C* unit to the offspring. A hybrid *Cc* or *cC* will contribute either a *C* or a *c*. The germ cells of three *C*-bearing plants inbred with the germ cells of the *cc* plant can produce twelve possible combinations of offspring. We can predict that eight of every twelve seeds will be yellow. Four of every twelve will be green. Once again, this turns out to be the case. Three of every four seeds, a ratio of 3:1, are yellow.

Thus it was clear to Mendel that inheritable characteristics are controlled by the combination of separate units of heredity. These hereditary units are contained in the reproductive cells, or gametes, of each parent. Alternate forms of the units (called *alleles*) often exist and are responsible for different forms of the inherited characteristic (height, for example, or the color of a flower petal). When they divide during meiosis, the gametes (sperm and ova) of each parent contribute half of their inheritance units to the offspring, one unit from each pair.

LAWS OF GENETIC INHERITANCE AND MUTATION

Mendel was hypothesizing the existence of what today are called genes, the physical unit of heredity. However, he never really discovered them. Nor did he use the name gene. That label was first applied to units of heredity by Dutch biologist Wilhelm Johannsen in 1909. Mendel may also have been unaware of the existence of chromosomes, the physical structures containing the genes of a particular species and individual. Chromosomes were first observed in the 1860s, about the time Mendel was writing up the results of his years of experiments. The term itself was first regularly used by Wilhelm von Waldeyer in 1888, although other researchers had used the word to describe the stained filament as early as 1873. *D*eoxyribonucleic acid, (DNA) was discovered in 1869 by the Swiss scientist Frederick Miescher. (RNA was also discovered in the late 19th century.) However, Mendel certainly did not know that DNA was the complex double-helix-shaped protein of which genes and chromosomes are made, and which is the very key to heredity and genetics. No one knew that for sure until 1953. We do know, however, of the existence of chromosomes, genes, and DNA. We now know the mechanisms for Mendel's empirically-based laws of inheritance. Those three laws are "The Law of Segregation," "The Law of Dominance," and "The Law of Independent Assortment," defined here:*

The Law of Segregation states that "as the gametes [mature male or female reproductive cells; sperm or ovum] form, the pairs of identical genes separate and do not influence each other."

The Law of Dominance states that some variations of genes will be "dominant" and others will be "recessive." If an individual inherits either a copy of a dominant and a recessive gene, or two copies of a dominant gene (one from each parent), the dominant gene(s) will be expressed. That is, the characteristic which the dominant gene causes will in fact be activated and present in that individual. Only if the individual inherits two copies of a recessive gene from both parents will that recessive gene be expressed.

The Law of Independent Assortment states that "traits controlled by different gene pairs (such as height and hair color) pass from parents to offspring independently of each other."

Mendel's work lay unnoticed for nearly 40 years. In 1900 it was rediscovered by Dutch botanist Hugo De Vries (1848–1935). De Vries and two other researchers had separately conducted their own series of experiments and discovered the basic laws of heredity. Before they submitted their results for publication, each of them had looked through the published

[*Definitions based on those in *Taber's Cyclopedic Medical Dictionary*, 14th Edition]

scientific literature to see what else had been written on the subject. After De Vries came across the published findings of Mendel, he and the others published the results of their experiments, but did not take credit for discovering the laws of heredity. Instead, they noted that their work served to confirm the validity of the work done by a Catholic monk more than 40 years earlier.

De Vries is not remembered as the man who discovered the laws of heredity, because he did not. He did, however, make another contribution to science that is all his own: the theory of *mutation*, an instance of transformation or change in a living creature. More specifically, a mutation is a *permanent* change in the genetic structure of a creature, in a gene or genes—a change that can be passed on to the creature's offspring. Such a change occurs in a gene (genes) in a reproductive cell of the creature, and that is why it can be passed on to future generations. A mutation in a gene inside any other cell of the creature (a skin cell, for example, or a muscle or nerve cell) will not be passed on to its offspring. Any physical or chemical changes caused by the mutation will only affect that specific individual.

Mutations in germ cells, however, affect that creature's descendants. The gene responsible for eye color, for example, may suddenly undergo a change in the germ cells of a person. Instead of causing brown eyes, the gene will now cause the eye color to be hazel. That gene is passed on to the next generation. Under certain circumstances, therefore, the child receiving the mutated gene will have hazel eyes instead of blue eyes. And that child in turn will eventually pass the hazel eye gene on to his or her children. Mutations are important in *evolution*, the process by which one species of plant or animal eventually changes into another new species.

The laws of genetic transmission and the process of mutation explain why children carry various physical characteristics of both their father and mother, or of other relatives. They also put teeth into Darwin's theory of evolution. The *fact* of evolution was attested to by the fossil evidence uncovered by geologists and paleontologists. The *how* of evolution, though, had been rather fuzzy. Now scientists could understand how the evolutionary process could work.

DRAWING THE OUTLINE

During the first half of the twentieth century, many important discoveries in genetics took place that essentially drew the rough outline for the genetic maps that would be created in the 1980s and 1990s.

For example, in 1905 the first genetic trait was located on a chromosome. It was sex itself. Nettie Stevens and Edmund Wilson of Columbia University discovered the existence of sex chromosomes. A pair of X chromosomes is found in women. Men, however, have only one X chromosome paired with a Y chromosome. The two are so named because of their shape as seen under a microscope.

One giant of genetics during the first part of the twentieth century was Thomas Hunt Morgan. He worked first at Columbia University. Later in his career he was one of many eminent scientists who migrated west to the tiny town of Pasadena, California, to turn an unknown college into the world-renowned California Institute of Technology, Caltech. Morgan concentrated on the genetics of the common fruit fly, *Drosophila melanogaster*. Fruit flies have only four chromosomes, and they are very large. They made excellent animals for genetics experiments. Morgan showed conclusively that genes are indeed on or in chromosomes. He also proved that genes must be in linear order within or upon chromosomes, and not merely jumbled around.

In 1926 Hermann Muller, a student of Hunt's, and Lewis Stadler independently discovered that X rays cause mutations in chromosomes and genes. That soon led to techniques for deliberately causing mutations in fruit flies.

In the 1940s and 1950s many researchers made slow but steady process in identifying the location of genes on chromosomes. The greatest breakthrough, of course, was the discovery of the double helix structure of DNA by James Watson and Francis Crick. Other researchers tried to meld together biology and mathematics into a science of "biomathematics." These first attempts failed. The mathematical, biological, and technological tools necessary were simply not available at the time.

However, by the 1960s the advances in biology and genetics were in place that would lead to a nexus, a point in the history of science and technology when something not previously possible became inevitable: The Genome Project. That nexus would be reached in 1985. The major players of the 1950s, 1960s and 1970s in the concatenation of discoveries and breakthroughs were James Watson, Francis Crick, and Maurice Wilkins; Paul Berg and David Jackson; Herbert Boyer and Robert Heming; Stanley Cohen and Annie Chang; Janet Mertz and Ronald Davis; and Walter Gilbert, Alan Maxam, and Frederick Sanger.

WATSON, CRICK, AND THE DOUBLE HELIX

The story of the discovery of the structure of DNA is so well known that it might seem to hardly bear retelling. Yet the tale is worth the hearing, if only in brief form, for the three-dimensional structure of DNA, along with its special combination of four chemical bases, is the key to its role as the container and sustainer of genetic information. Its arrangement was discovered in 1953 by James Watson and Francis Crick.

Watson received his PhD in 1950 at Indiana University. He had been working on *phages*, viruses that infect and kill bacteria, and went to Copenhagen on a fellowship to continue that work. However, his interests then shifted. He decided he wanted to study molecular structure, and to learn about X-ray diffraction techniques to do that. In the fall of 1951 he began working at the Cavendish Laboratory at Cambridge, and ended up meeting and working with Francis Crick.

Crick was Watson's senior by ten years. He was a physicist who became interested in biology and changed fields. When he and Watson met in October 1951, Crick was working on DNA's three dimensional structure as part of his PhD thesis. Later, he ended up becoming one of the world's premier authorities on molecular biology.

Setting the stage for Watson and Crick's discovery was the earlier work of three eminent scientists. In 1944 physician and microbiologist Oswald Avery was doing research on a substance in the pneumococcus bacteria which had the property of a gene. (The pneumococcus bacteria causes pneumonia.) Avery proved conclusively that this gene-like substance was DNA. Clearly that molecule was a key to the process of genetic inheritance. Not everyone accepted Avery's conclusion, though. Many scientists felt that some protein was the genetic material. Avery's work did not absolutely exclude that hypothesis. So many remained skeptical of his assertions about DNA.

Nine years later, in 1950, chemist Erwin Chargaff discovered that the amount of adenine in DNA always equalled the amount of thymine, and that the amount of guanine always equalled the amount of cytosine. He was sure that these intriguing proportions said something about the physical structure of DNA, but he didn't know exactly what. Nor did Chargaff pursue his intuition.

A third scientist who played a major role in leading to Watson's and Crick's DNA breakthrough was the American chemist Linus Pauling. Pauling is without doubt *the* 20th century wizard of structural chemistry. He is best-known today for his enduring championship of vitamin C as an effective medicine for the common cold and cancer. In the 1960s he received the Nobel Peace Prize for his tireless work in behalf of nuclear disarmament. But his *first* Nobel medal was for chemistry. During the 1940s and 1950s he had no peer in determining the three-dimensional shapes of molecules. This, of course, was what Crick and Watson wanted to find out about DNA. Pauling's contribution was two-fold. First, he pioneered the use of model-building. Three-dimensional structures made of wood, cardboard or metal physically incorporated the known data about a molecular structure. If the model fit together well, the scientist would reasonably assume that it was a good representation of the real thing. Of course, that would still have to be proved.

One way to prove it was with Pauling's other significant contribution: X-ray crystallography. Pauling led the way in interpreting the patterns on a photographic plate made as X rays passed through crystals of a particular substance. Two researchers working in the same lab as Watson and Crick used the technique to show that helical structures were an important part of the hemoglobin molecule. Francis Crick himself used crystallography, and knew that X-ray analysis would be important in uncovering the structure of DNA. Jim Watson, in fact, came to England to learn X-ray crystallography from Crick. When Pauling made known his interest in learning DNA's structure, the two researchers' competitive drive was kicked into high gear.

Two other researchers also played an important role in the discovery of DNA's structure. One of them later shared the Nobel Prize with Watson and Crick. That was Maurice Wilkins, a physicist at King's College in London who had begun studying DNA's structure using X-ray crystallography in 1950. The other was Wilkins' young associate, Rosalind Franklin. She was a methodical experimenter and careful examiner and interpreter of the X-ray photos Wilkins produced. She was very cautious in her work, and in her comments about it. The rapport between Franklin and Wilkins was formal and sometimes strained. Franklin was reluctant to share her findings and ideas with Wilkins. In fact, people who knew her later remarked that Franklin's caution often crossed the border into secretiveness. She was always very reluctant to talk with other researchers about her work. Perhaps this arose from a fear that others might steal her data and claim it as their own. Perhaps it was a result of being one of a very small number of women in what was in the 1950s an almost exclusively male preserve—science in general and molecular biology in particular.

By contrast James Watson and Francis Crick—two very different people physically and temperamentally—got along very well together, with much give-and-take and open, easy conversation. In hindsight, Wilkins and Franklin could have uncovered DNA's structure more than a year before Crick and Watson did; but the prize, and the Prize, went to the brash young American and the older, outspoken Britisher.

In November 1951, a month after Crick and Watson began working together at the Cavendish, Rosalind Franklin at King's College obtained X-ray diffraction images that revealed a key to DNA's structure. But she didn't recognize it. Meanwhile, Watson and Crick thought they had enough data to begin building a model of DNA. Franklin and Wilkins came to visit them, saw their model—and shot it down in flames. Their criticism was so devastating that Crick and Watson's superior, Sir Lawrence Bragg, told them to stop their work on DNA. (There was also some politics involved in Bragg's order. He didn't want to step on Wilkins' toes.)

Six months later, Franklin obtained still more X-ray diffraction photos of DNA that indicated it had a helix structure. Still she missed seeing the evidence. At about the same time, mathematician John Griffith, a friend of Crick's, produced some calculations showing that adenine would attract thymine and guanine, cytosine. Shortly afterward, Erwin Chargaff visited Cambridge and met Crick and Watson. He told them of his discovery that the amount of adenine and thymine as well as the amount of guanine and cytosine, were always the same in DNA. Crick was struck by the one-to-one ratios of these chemical components. He realized that this could be the basis for a molecule being able to duplicate itself.

Crick and Watson's motivational turning point came in November 1952. Linus Pauling announced that he was interested in DNA, too, and planned on deciphering its three-dimensional structure. In January 1953 he sent Watson and Crick a copy of a manuscript laying out his proposal for DNA's structure. The two men had been on tenterhooks. Pauling was a "top gun" in molecular biology and chemistry, and they feared he would surely beat them out in the race to uncover DNA's structure. When they read Pauling's manuscript, they breathed a sigh of relief. Some of his X ray data was obviously in error. Pauling's model was wrong.

Lawrence Bragg now had a a change of heart, and again for mainly political reasons. Since Pauling was not close to the structure, and apparently Wilkins and Franklin were going nowhere, his boys Crick and Watson just might be able to grab the brass ring. He gave them permission to drop their other work and concentrate on DNA's structure. Watson got a chance at this point to look at some of the X-ray diffraction photos made by Rosalind Franklin and Maurice Wilkins. His analysis of the patterns convinced Watson that the images showed DNA to be helical in structure, with two chains. When Crick saw Franklin's X ray images made in November 1951, he deduced that DNA must be a double helix with its chains running in opposite directions. The base-pair matching of A with T and C with G would make it so, he realized.

On February 28, 1953, Watson made some cardboard cutouts of the four bases, shaping them according to the suggestions of a former colleague of Pauling's named Jerry Donohoue. The supposedly complementary pairs fit together. Watson then attached them to a model of the backbone sugar-phosphate chains, with the bases pointing inward. They fit perfectly when the two chains' order of bases ran in opposite directions. At the beginning of March 1953 they built their final three-dimensional model of DNA, and sent out invitations to their colleagues to critique it. Nearly everyone who saw it was struck by a sense of its "rightness." The model's very

This is how DNA actually looks, magnified a million times. This image is the first photo of individual strands of unaltered DNA to show the double helix structure. The image was made by researchers at the Lawrence Livermore National Laboratory, the Lawrence Berkeley Laboratory, and the University of California at Berkeley. The scientists used a scanning tunneling microscope, a device that can record structural details of surfaces as small as a single atom.
—Photo courtesy of Lawrence Livermore National Laboratory.

beauty argued for its correctness. In April, Linus Pauling arrived from America to take a look at their model.

He liked it.

On April 25, 1953, Watson's and Crick's three-page paper appeared in *Nature* magazine. It was short and understated. And almost no one was fooled by that at all. In 1962, Watson, Crick, and Wilkins won the Nobel Prize for Medicine for their discovery, one of the greatest scientific breakthroughs of this or any other century. The Watson-Crick model of DNA's three-

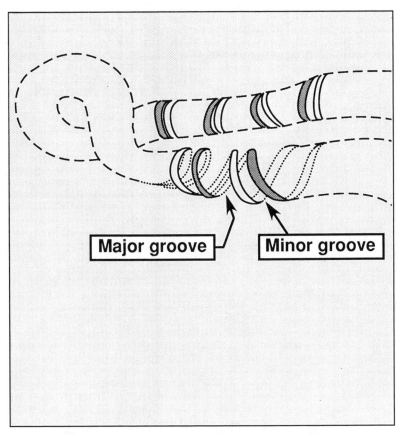

This drawing clarifies some of the structure visible in the DNA photo, including twisting strands which make up the double helix structure of the molecule.—Illustration courtesy of Lawrence Livermore National Laboratory.

dimensional structure has, with minor modifications, has proven to be correct.

Watson and Crick conclusively showed that *DNA*, or *deoxyribonucleic acid*, is the key to heredity. The molecule has the particularly elegant shape of a double helix, and looks like a ladder whose outer rails are twisted about each other. The rails of the DNA ladder are made of units of sugar and phosphate atoms. The ladder's rungs are made of pairs of molecules called bases linked to the units of the outer skeleton. The base-

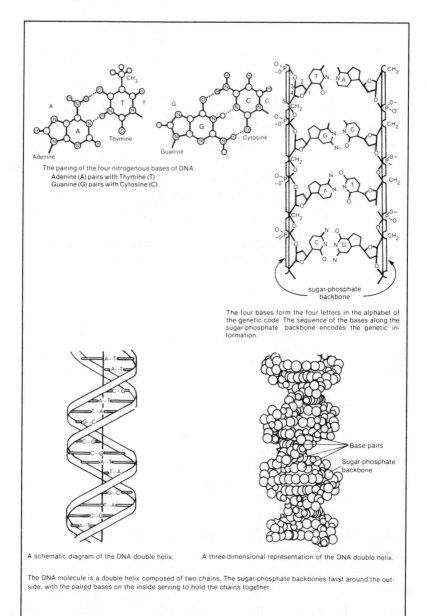

The pairing of the four nitrogenous bases of DNA:
Adenine (A) pairs with Thymine (T)
Guanine (G) pairs with Cytosine (C)

sugar-phosphate backbone

The four bases form the four letters in the alphabet of the genetic code. The *sequence* of the bases along the sugar-phosphate backbone encodes the genetic information.

A schematic diagram of the DNA double helix.   A three-dimensional representation of the DNA double helix.

The DNA molecule is a double helix composed of two chains. The sugar-phosphate backbones twist around the outside, with the paired bases on the inside serving to hold the chains together.

The Structure of DNA.—Reprinted from: U.S. Congress, Office of Technology Assessment, "Technologies for Detecting Heritable Mutations in Human Beings" OTA-H-298 (Washington, D.C.: U.S. Government Printing Office, September 1986.)

skeletal unit combination is called a *nucleotide*. The four bases in DNA are *adenine, guanine, thymine,* and *cytosine.* They are usually abbreviated to A, G, T, and C. Because of their chemical properties, T can link up only with A, and G only with C (and vice-versa). Each pair of bases, A-T, T-A, C-G, or G-C is called a base pair. The rungs of the DNA molecular ladder are either A-Ts or C-Gs.

However, along the inside of any strand or chain of DNA nucleotides, the As, Ts, Cs and Gs can be in any order what-soever. For example, one segment of DNA could be

TACGGCGGCTTCTTG.

The four bases can be in any order at all along a chain of DNA; it's one side of the double helix. However, if we know the order of the four bases on one side of the double helix ladder, than we can figure out the sequence on the other side. This is called *complementary base pairing.* One chain always acts as the template for the other. One segment of a typical DNA molecule might be this:

T A C G G C G G C T T C T T G
| | | | | | | | | | | | | | |
A T G C C G C C G A A G A A C

CRACKING THE CODE

The genetic code is a language with just four letters and sixty-four words. Each word is three letters long and is called a *codon*. Sixty-one codons are the codes for twenty *amino acids*, the chemical compounds which are the building blocks of proteins. One codon act as a "start" signal, signifying the beginning of a protein chain. Three codons are "stop" signals, signaling the end of a protein chain. A *gene* is a series of codons containing the instructions for making a protein. But when geneticists talk about the genetic alphabet, they are not actually speaking about DNA. The four letters of the genetic alphabet are the nucleotide bases that make up *RNA*, or *ribonucleic acid*. This is a single-strand molecule very similar to DNA which carries the original coded message from DNA to the parts of the cell which turn the code into amino acids. The letters of the RNA genetic code are the same as those in DNA, with one exception. Uracil substitutes for thymine in RNA. Uracil was vital to the experiment which cracked the genetic code. It was carried out by Marshall Nierenberg, Severo Ochoa, and Har Gobind Khorana, and their work was a climax to a series of breakthroughs that followed the world of Crick and Watson.

One important breakthrough was made by Charles Yanofsky at Stanford University and Sydney Brenner at the Cavendish Laboratory in England. These two researchers, working independently, showed that there is a clear linear relationship between the arrangment of bases in the DNA code and the amino acids in a protein. In the process, they also provided experimental evidence for a theory suggested by Francis Crick that it takes three DNA bases to code for one amino acid. Crick's belief was based on simple math. If there is a linear relationship between DNA bases and the protein for which it codes, then some specific number of the four bases which constitute DNA must determine which of the 20 amino acids are located in any particular place in a protein. Groups of four bases cannot *unambiguously* code for 20 amino acids—there are 256 possible combinations, far too many. Pairs made from two of the four bases make up only sixteen possible combinations—too few. However, there are 64 possible *triplets* of the four nucleotides. That's enough to code for 20 amino acids, plus several back-ups for each. Yanofsky's work in particular provided concrete evidence for Crick's mathematical suggestion. He had found a linear relationship between a gene made of about a thousand nucleotides, and a protein about 300 amino acids long. The proportion is about 3.3 to one, very close to Crick's theoretical 3 to 1 ratio of nucleotides to protein.

The next breakthrough was the result of a series of experiments over several years by researchers in France, England, and the United States. They included François Jacob and Jacques Monod in France, Francis Crick and Sydney Brenner in England, and Mahlon Hoagland, Matthew Meselson, Arthur Pardee, and Paul Zamecnik in the United States. Separately and together in various research teams, they accumulated convincing evidence that RNA is crucial to the functioning of the genetic code. RNA, they found, carries the information contained in DNA to a location in a cell that might best be called a protein factory. Other kinds of RNA then help to translate the code into amino acids and proteins.

However, the actual code itself was still unbroken. That's where Nirenberg, Ochoa, and Khorana made their Nobel-winning contribution. The key was uracil, the fourth nucleotide in RNA. In 1961, Nierenberg and Johann Matthei were young biochemists at the National Institutes of Health in Bethesda, Maryland. They wanted to find some direct evidence for whatever kind of nucleotide templates coded for amino acids and proteins. The lab next to theirs happened to be involved in the creation of artificial RNA, strings of nucleotides connected by means of a newly discovered enzyme. One such artificial RNA was called PolyU. It was nothing more than a long sequence of uracils—UUUUUUUUU. . . . Nierenberg and Matthei took some PolyU and added it to a test tube containing the protein-making machinery they had previously removed from a living cell. Then they added a mixture of the 20 amino acids. The result was an utter surprise. PolyU was taking only one amino acid out of the mixture: phenylalanine. It created an artificial "protein" called polyphenylalanine—phenylalanine repeating itself over and over and over. . . . If Crick's coding theory was indeed right, and triplets of nucleotides coded for amino acids, then the two men had just cracked the code for the amino acid phenylalanine: UUU. This triplet was eventually called a codon.

Severo Ochoa (who had invented the method of making artificial RNA) and his colleagues at New York University quickly jumped into the fray. Between 1961 and 1966 the codes for every amino acid were uncovered. In the process it became clear that—as Crick's hypothesis had suggested—several amino acids had two or more triplet combinations.

However, a further question remained unanswered: which is the proper arrangement of nucleotide letters for a particular triplet? The CAA codon is not the same as ACA. Which is the correct spelling for a particular amino acid? The answer came in 1964, from Nirenberg and another up and coming researcher named Phil Leder. They found the correct spelling for the amino acid valine. The codon for valine was either GUU, UGU, or UUG. The two researchers found that only GUU correctly coded for valine. UGU, they later found, was the codon for cysteine and UUG for leucine. Their work was confirmed and extended by Har Gobind Khorana at the University of Wisconsin.

By the end of 1968, the entire genetic code had been broken, for all 64 amino acids. Argenine, leucine, and serine each have six different codons. Five amino acids have four codons each: alanine, glycine, proline, threonine, and valine. Isoleucine is coded for by three different codons. Methionine and tryptophan each have only one. Methionine's codon also can act as a "start" signal. The remaining nine amino acids have two codons each.

Let's look at how codons are read in the following strand of RNA:

UACGGCGGCUUCUUG

For example, the first codon in the RNA molecule above, UAC, is one of the two codons for the amino acid tyrosine. The second and third codons, GGC, is one of the four genetic codes for glycine. UUC is one of the codons for phenylalanine. The fifth codon, UUG, is one of the six "words" that spells out leucine. The entire sequence of 15 nucleotides is the code for

Codon	Amino Acid	Codon	Amino Acid	Codon	Amino Acid	Codon	Amino Acid
UUU	Phenylalanine	UCU	Serine	UAU	Tyrosine	UGU	Cysteine
UUC	Phenylalanine	UCC	Serine	UAC	Tyrosine	UGC	Cysteine
UUA	Leucine	UCA	Serine	UAA	stop	UGA	stop
UUG	Leucine	UCG	Serine	UAG	stop	UGG	Tryptophan
CUU	Leucine	CCU	Proline	CAU	Histidine	CGU	Arginine
CUC	Leucine	CCC	Proline	CAC	Histidine	CGC	Arginine
CUA	Leucine	CCA	Proline	CAA	Glutamine	CGA	Arginine
CUG	Leucine	CCG	Proline	CAG	Glutamine	CGG	Arginine
AUU	Isoleucine	ACU	Threonine	AAU	Asparagine	AGU	Serine
AUC	Isoleucine	ACC	Threonine	AAC	Asparagine	AGC	Serine
AUA	Isoleucine	ACA	Threonine	AAA	Lysine	AGA	Agrinine
AUG	Methionine (start)	ACG	Threonine	AAG	Lysine	AGG	Arginine
GUU	Valine	GCU	Alanine	GAU	Aspartic acid	GGU	Glycine
GUC	Valine	GCC	Alanine	GAC	Aspartic acid	GGC	Glycine
GUA	Valine	GCA	Alanine	GAA	Glutamic acid	GGA	Glycine
GUG	Valine	GCG	Alanine	GAG	Glutamic acid	GGG	Glycine

Each codon, or triplet of nucleotides in RNA, codes for an amino acid (AA). Twenty different amino acids are produced from a total of 64 different RNA codons, but some amino acids are specified by more than one codon (e.g., phenylalanine is specified by UUU and UUC), In addition, one codon (AUG) specifies the "start" of a protein, and 3 codons (UAA, UAG, and UGA) specify termination of a protein. Mutations in the nucleotide sequence can change the resulting protein structure if the mutation alters the amino acid specified by a triplet codon or if it alters the reading frame by deleting or adding a nucleotide.
U—uracil (thymine)
C—cytosine
A—adenine
G—guanine

The Genetic Code.—Reprinted from: U.S. Congress, Office of Technology Assessment, "Mapping Our Genes—The Genome Project: How Big, How Fast?" OTA-BA-373 (Washington, D.C.: U.S. Government Printing Office, April 1988.)

a short protein (called a peptide) known as *leucine-enkephalin*, one of the endorphin family of proteins in the brain and body.

LEVELS OF INFORMATION

Genes are the basic units of heredity in living creatures. Different genes carry the instructions for different inheritable characteristics. Genes, in turn, are grouped together in distinctive structures inside the nucleus of a cell, called *chromosomes*. The human genetic code is contained in 46 individual chromosomes grouped into 23 pairs. Twenty-two of the chromosome pairs are called *autosomes* and are numbered 1 through 22. The remaining two chromosomes are the *sex chromosomes*, called X and Y.

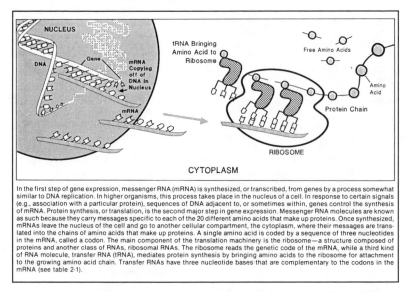

In the first step of gene expression, messenger RNA (mRNA) is synthesized, or transcribed, from genes by a process somewhat similar to DNA replication. In higher organisms, this process takes place in the nucleus of a cell. In response to certain signals (e.g., association with a particular protein), sequences of DNA adjacent to, or sometimes within, genes control the synthesis of mRNA. Protein synthesis, or translation, is the second major step in gene expression. Messenger RNA molecules are known as such because they carry messages specific to each of the 20 different amino acids that make up proteins. Once synthesized, mRNAs leave the nucleus of the cell and go to another cellular compartment, the cytoplasm, where their messages are translated into the chains of amino acids that make up proteins. A single amino acid is coded by a sequence of three nucleotides in the mRNA, called a codon. The main component of the translation machinery is the ribosome—a structure composed of proteins and another class of RNAs, ribosomal RNAs. The ribosome reads the genetic code of the mRNA, while a third kind of RNA molecule, transfer RNA (tRNA), mediates protein synthesis by bringing amino acids to the ribosome for attachment to the growing amino acid chain. Transfer RNAs have three nucleotide bases that are complementary to the codons in the mRNA (see table 2-1).

Gene Expression.—Reprinted from: U.S. Congress, Office of Technology Assessment, "Mapping Our Genes—The Genome Project: How Big, How Fast?" OTA-BA-373 (Washington, D.C.: U.S. Government Printing Office, April 1988.)

■ Some DNA sequences regulate gene expression—that is, they somehow tell the genes when to turn on and turn off. This is absolutely vital to the functioning of any living creature. Every cell (except for germ cells, sperm and ova, which carry one-half the set of chromosome pairs) contains the entire genome, a complete set of chromosomes. But only certain genes are active in specific cells. That's why bone cells are different from skin cells, nerve cells from blood cells, the immune system's T cells from its B cells.

■ The genome also contains *structural DNA*, which is needed to package chromosomes and to create their physical structure.

■ Then there is so-called "junk DNA." This is DNA that seems to code for nothing at all. It is also called "silent DNA." Of course, many people make a distinction between junk and garbage. Garbage gets thrown out, while junk is saved—who knows, some of that junk might be useful some day. In the same way, not every researcher thinks that junk DNA is necessarily garbage. It is present still in the genome. It could very well be that these long segments of DNA do something, but no one knows yet what it is.

Cells make many thousands of different proteins. Each cell (except for sex cells) contains a full complement of genes—perhaps 100,000—in its chromosomes. To make a particular protein, the cell must copy the correct information from the correct gene. The process by which the DNA codes are translated into proteins (and ultimately into a human being) has two major steps, called *transcription* and *translation*. Both depend on the actions of RNA. RNA comes in several varieties. They include *messenger RNA*, or mRNA, *transfer RNA*, or tRNA, and *ribosomal RNA* (rRNA).

FROM CODE TO PROTEIN

Transcription of the DNA code starts with the action of a chemical called *RNA polymerase*. This enzyme "unzips" the two intertwined DNA nucleotide chains in a short region of the molecule. The region is about ten to twenty base pairs long. The RNA polymerase slides along the DNA molecule, unzipping that small region as it moves. As it does, it creates an RNA chain out of free-floating nucleotides. The single RNA chain is a copy of one of the DNA chains. The RNA's nucleotides are in an order determined by complementary base pairing to the DNA strand. When the RNA polymerase sees a G in the DNA strand, for example, it adds a C to the RNA strand it is making. If it sees a C, it adds a G. If it sees a T it adds an A, but if it see an A, it adds a U (for uracil), since RNA uses uracil instead of thymine. For the DNA molecule mentioned earlier, for example, RNA polymerase would create an RNA chain for the upper DNA chain by reading the lower chain and adding or substituting the appropriate bases. The DNA is:

```
T A C G G C G G C T T C T T G
| | | | | | | | | | | | | | |
A T G C C G C C G A A G A A C
```

The RNA transcription for the upper chain would be

UACGGCGGCUUCUUG

This RNA molecule is messenger RNA, or mRNA. Its function is just that: It carries the information originally encoded in a stretch of DNA out of the nucleus to a location where it will be turned into a protein. These "protein factories" are the *ribosomes*, ball-like structures made of ribosomal RNA and several specialized proteins. It is in the ribosomes that the four-letter code of DNA/RNA is turned into the 20-letter code of amino acids and proteins. Here's how it works:

Floating around in the cellular region near the ribosome are amino acids and various strands of transfer RNA. A special enzyme called *aminoacyl-tRNA synthetase* connects specific amino acids to specific pieces of tRNA. Twenty types of tRNA exist, one for each amino acid. Each tRNA strand has at one end a three-nucleotide sequence called an *anticodon*. The amino acid hooks onto the tRNA strand at the opposite end. Thus, the anticodon at one end of the tRNA always corresponds to the amino acid at the other.

The translation process begins when the messenger RNA from the nucleus attaches itself to the ribosome. The connection takes place at a point on the mRNA strand near its start codon. The mRNA is held in the ribosome somewhat like a slide tape is held in a projector. It gets pulled through the projector frame by frame. So does the mRNA through the ribosome.

Along comes a tRNA strand with its attached amino acid. The tRNA's anticodon is complementary to the first codon on the mRNA. The tRNA and mRNA lock together and the codons form base pairs. So, for example, the RNA start codon is AUG. Only a tRNA strand with an anticodon complementary to this (UAC) will fit onto that part of the mRNA. The next codon on

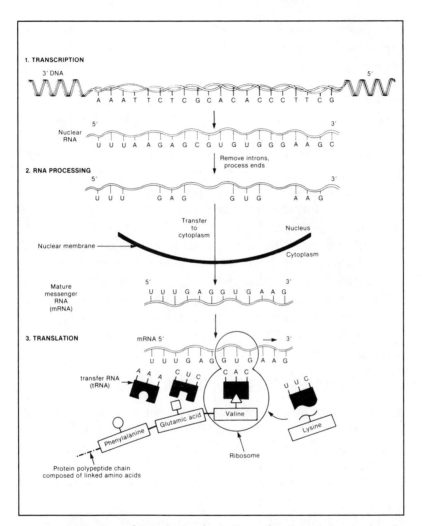

A Diagram of Protein Synthesis From the Genetic Instructions in DNA. In the transcription process, a segment of DNA is transcribed into an RNA molecule. The introns are then snipped out, and the mature messenger RNA (mRNA) molecule is moved out of the cell's nucleus to a ribosome. In the translation process, transfer RNA (tRNA) molecules help translate the genetic code in the mRNA into amino acids, which are linked together to form a peptide or protein.—Reprinted from: U.S. Congress, Office of Technology Assessment, "Technologies for Detecting Heritable Mutations in Human Beings" OTA-H-298 (Washington, D.C.: U.S. Government Printing Office, September 1986.)

the mRNA is now locked into place, and it is quickly recognized by a tRNA molecule with the correct anticodon. The second tRNA locks on. Specialized proteins in the ribosome now join together the two amino acids that are on the other ends of the tRNA molecules. The first amino acid cuts loose from its tRNA, and that strand separates from the mRNA and floats off to begin its work anew.

The mRNA now begins sliding through the ribosome like a magnetic tape past the playing head of a tape recorder. Successive tRNA molecules—with their appropriate anticodons and attached amino acids—hook onto the mRNA. Their amino acids are thus correctly positioned, linked together, and cut away from their tRNA carrier strands. Finally a stop codon in the mRNA reaches its correct position in the ribosome. The translation process stops, and the entire amino acid chain is released into the cell. The chain spontaneously folds itself into the correct three-dimensional shape of the completed protein. The newly created protein is now ready to begin its work in the cell.

FROM DNA TO DNA TO DNA

The essence of inheritance is the ability of parents to pass on genetically controlled characteristics to their children. This means that DNA must be able to replicate itself. More than that, it must be chemically stable and able to replicate itself faithfully, with no mistakes. In this way there is genetic continuity from one generation to the next.

DNA is quite stable chemically, and it almost always duplicates itself with no errors. Sometimes, though, mistakes happen. Errors do creep in. They may be caused by chemicals or by radiation, or by an imperfect replication process. These changes in genetic coding from one generation to the next are called *mutations*. Mutations drive evolution, the process by which one species changes into another over time. In general, however, DNA replication is stable enough to ensure that mutations are rare.

The process by which the information in DNA is passed on to new generations begins when the two strands of a DNA

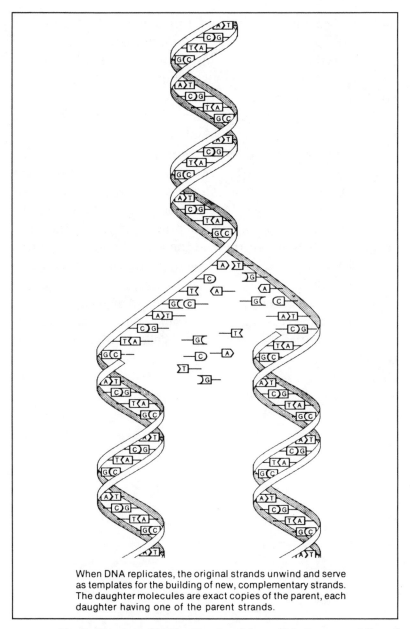

When DNA replicates, the original strands unwind and serve
as templates for the building of new, complementary strands.
The daughter molecules are exact copies of the parent, each
daughter having one of the parent strands.

The Replication of DNA.—Reprinted from: U.S. Congress, Office of
Technology Assessment, "Mapping Our Genes—The Genome Proj-
ect: How Big, How Fast?" OTA-BA-373 (Washington, D.C.: U.S.
Government Printing Office, April 1988.)

molecule separate. This makes it possible for each strand to act as a template for the creation of a new strand. A complex series of actions then takes place, involving many different enzymes and other compounds within the cell. Eventually individual nucleotide units within the cell attach to the two strands. They follow the complementary base pair rule. Adenine nucleotides attach to thymines on the strand, guanines to cytosines, and so on. As they join to the strand, they also connect to one another. The result is the creation of two new DNA molecules—called *daughter molecules*—where once there was one. The two new DNA molecules now move to different parts of the cell nucleus, and the process of cell division and reproduction begins. In the end there are two new daughter cells, each containing the same genetic information as was in the orginal cell.

This is just a general sketch of how DNA replication works. The actual process is somewhat more complicated. Nevertheless, it is essentially the same in all cells, in all life forms on this planet.

3

Maps, Splices, and a "Moratorium"

The only solid piece of scientific truth
about which I feel totally confident
is that we are profoundly ignorant about nature.

LEWIS THOMAS
The Lives of a Cell

ONCE NIRENBERG and others had found the key to deciphering the genetic code, it was broken fairly quickly. That code proved to be the same in all creatures, but most geneticists in the 1960s and early 1970s were focusing their research on the simplest life-forms around: *prokaryotes*. They are better known to most people as bacteria. The deadly salmonella "bug" is one. So is *E. coli*, a bacterium found in people's intestines. Prokaryotes have a simple genetic system and physical structure, making them easy to study and understand. However, they *are* simple, and in many ways unlike more complex living creatures made of more complicated cells called *eukaryotes*. That includes humans.

Meanwhile, other researchers were deciphering still another "code"—the language of genes and chromosomes. Slowly but surely, scientists were learning the locations of specific genes on specific chromosomes, and learning which genes caused specific genetic diseases. Progress was slow but sure. By the early 1960s, in fact, enough had been learned to write a book. And that book would grow enormously. Its author—or to be more precise, its compiler and editor—is Victor McKusick.

VICTOR McKUSICK AND MIM

Victor McKusick of the Johns Hopkins University in Baltimore, Maryland, is a very tall man, with white hair and a soft melodic voice. He talks quickly, and often stumbles over his words as he speaks. He wears a white lab coat as he works in his office. He seems the archetypal scientist. He is more than that. Victor McKusick is a giant figure in modern genetics. And as he approaches his eighth decade he shows no signs of slowing down.

McKusick's towering achievement is a book that has never been completed. It is called *Mendelian Inheritance in Man* [sic], or MIM. MIM is essentially an encyclopedic listing of human gene loci, of the locations of genes on the 23 pairs of chromosomes that comprise the human genome. The information in MIM comes from researchers around the globe. Though published in the United States, MIM is a global volume. The first edition, MIM 1, was published in 1966. MIM 8, the eighth edition, was released 22 years later, in July 1988. The number of entries has skyrocketed in those two decades. MIM has become the central focus, the touchstone, of McKusick's career. It is also an ongoing record of the growth of our understanding of the human genome.

The first human genetic trait to be identified and linked to a specific chromosome was color-blindness. E.B. Wilson, the famous psychologist at Columbia University, deduced in 1911 that it was on the X chromosome. Wilson came to this conclusion by studying the characteristic pedigree or inheritance pattern of color blindness. After that, a fair number of conditions were deduced to be X-linked because of their pedigree patterns. It is not all that difficult a task, even in the years before genetic engineering. The X chromosome is the sex chromosomes that both men and women have. This makes it fairly easy to follow inheritance patterns of genetic diseases linked to maleness. Color-blindness, for example, is a hereditary condition that occurs almost exclusively in males. The most common form of color blindness is a difficulty in distinguishing red from green. Those who are completely red-green color blind see both colors as yellow. Totally color-blind people see only white, black, and shades of gray.

Two other X-linked conditions whose genes were mapped during the years before the development of genetic engineering were classic hemophilia A and Lesch-Nyhan Syndrome. In 1939 Julia Bell and John Burton Sanderson Haldane, the famous geneticist and popularizer of science, showed that hemophilia A is genetically linked to color-blindness. It is therefore located on the X chromosome near that gene. Hemophilia A is a disorder in which the clotting ability of the blood is impaired. A person with hemophilia bleeds easily and copiously from the slightest cut. Perhaps even more serious are bruises, which cause excessive or uncontrolled internal bleeding. The immediate cause of hemophilia is the absence of one or more of the molecular compounds in the blood, called clotting factors. That absence is caused by a defect in a gene on the X chromosome. Hemophilia is transmitted through females, but usually affects only males. Lesch-Nyhan syndrome is a tragic genetic illness characterized by mental retardation, aggressive behavior, self-mutilation such as compulsive biting of the fingers, and eventual kidney failure. It occurs only in males. The syndrome was first formally described by American pediatricians Michael Lesch and William Nyhan, Jr. in 1964, and was mapped to the X chromosome a year later by the American researcher Dick Hoefnagel.

By 1968 researchers had found and mapped 68 X-chromosome-linked genes. Sixty-eight genes mapped in fifty-seven years is not a lot. However, through the mid-1960s the technology for identifying and mapping genes and chromosomes was very primitive. It wasn't until 1956 that scientists even knew for sure that there were 46 human chromosomes in 23 pairs. In 1968 the first gene was located on an autosome, or non-sex chromosome. Roger Donahue mapped a particular blood group called the Duffy blood group to chromosome 1. Following in the classic scientific tradition of self-experimentation, Donahue studied the pedigree patterns of his own family in order to make the linkage. By 1971 three other autosomal genes had been mapped. And then the field took off.

The reason was the development of *somatic cell hybridization* or SCH, first demonstrated in 1967. This is a technique which involves fusing ("hybridizing") tumor cells of humans with those of another species, such as mice. This is done using certain chemicals or an electric field. A virus called the Sendai virus also helps fuse the cells. At first the fused cells contain

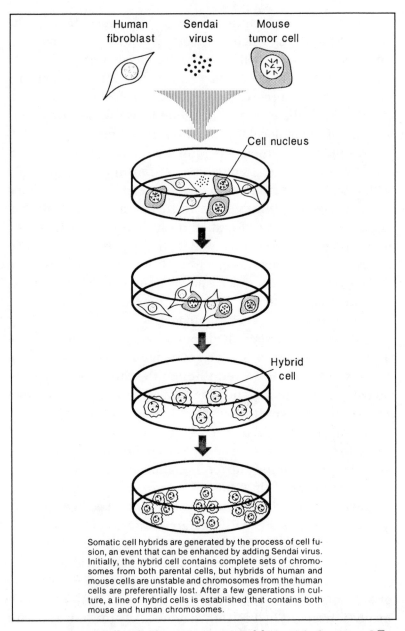

Somatic cell hybrids are generated by the process of cell fusion, an event that can be enhanced by adding Sendai virus. Initially, the hybrid cell contains complete sets of chromosomes from both parental cells, but hybrids of human and mouse cells are unstable and chromosomes from the human cells are preferentially lost. After a few generations in culture, a line of hybrid cells is established that contains both mouse and human chromosomes.

Somatic Cell Hybridization.—Reprinted from: U.S. Congress. Office of Technology Assessment, ''Mapping Our Genes—The Genome Project: How Big, How Fast?'' OTA-BA-373 (Washington, D.C.: U.S. Government Printing Office, April 1988.)

a complete set of chromosomes from the human and rodent parent cells. However, hybrid cells are unstable, and they begin losing chromosomes. The hybrid cells grow and divide, losing chromosomes with each generation until they finally stabilize. The human cells are more easily lost than the rodent chromosomes. The hybrid cell lines usually end up containing eight to twelve of the 46 human chromosomes, in addition to the remaining rodent chromosomes.

It is at this point that SCH becomes useful for mapping the location of genes on chromosomes. Researchers take a large set of somatic cell hybrids containing different combinations of chromosomes and compare the presence or absence of a specific human *chromosome* with the presence or absence of a particular *gene*. This is done by detecting a protein produced by the hybrid cell, and associating it with the chromosome *found only in that particular hybrid cell* line. Suppose protein "Z" is found only in a somatic cell hybrid line containing human chromosome 5. None of the other cell lines has human chromosome 5, so it is clear that the gene which codes for protein "Z" is on human chromosome 5. Since the development of SCH, it has become possible to create cell lines containing only one human chromosome. Hybrid cell lines containing single copies of human chromosomes 7, 16, 17, 19, X and Y are now used by researchers. They have made it much easier to create low-resolution physical maps of genes on those chromosomes.

In the early 1980s the molecular genetics techniques of genetic engineering started becoming available to many researchers. It marked another spurt in the rate of gene-mapping. Now it is increasing very quickly, indeed. It is increasing so fast, in fact, that Victor McKusick now uses an online version of MIM to keep continual track of the genes being linked to particular chromosomes. MIM has increased in size over the years. MIM 8 was printed on ultra-thin Bible paper. The computer files were closed in March of 1988, and the book was available just four months later. It was totally composed and typeset from the computer file tapes.

FROM X TO MIM

MIM was born from the self-admitted compulsive nature of Victor McKusick's search for knowledge. He calls himself an "encyclopedist," in the tradition of the great encyclopedist scientists of the seventeenth and eighteenth centuries such as Isaac Newton, Rene Descartes, and Benjamin Franklin. What McKusick wants to have is the whole picture of whatever it is he is studying. In 1962 it was the X chromosome.

McKusick's research goal at that time was to study the human X chromosome from every point of view. McKusick asked himself the question: "What information is contained in the X chromosome?" A partial answer can be supplied by simply listing all the traits known to be X-linked, such as hemophilia and color-blindness. A list of these diseases and conditions, says McKusick, is like a photographic negative from which a positive picture can be made. That picture will be of the genetic products of the X chromosome. So McKusick made a complete catalog of the X-chromosome-linked traits. He published that catalog in 1962. Another outcome of this work was his 1964 book, *On the X Chromosomal Bands*.

Then McKusick began branching out. He started studying the genetics of the Amish people, who live mainly in southeastern Pennsylvania. Because Amish rarely marry outside their own religion, McKusick thought they would tend to have more recessive genetic traits than the population at large. McKusick wanted to catalog those as well. One thing led to another, driven by his great desire to be as comprehensive as possible. Not too surprisingly, he followed this by cataloging all the known *dominant* genetic traits.

The first edition of MIM, based on his catalogs of X-linked traits, recessive traits, and dominant traits, was published in 1966. Both it and the next two editions were photo-offset

Victor McCusick has been a pioneer in the task of finding and listing genes responsible for or associated with thousands of diseases and genetic conditions.— Photo courtesy of Johns Hopkins University.

Paul Berg won the Nobel Prize for his breakthrough experiments in genetic engineering, was at the center of the Asilomar "moratorium," and is a persistent proponent of the Genome Project.

directly from computer tapes, and so were printed in upper case. Beginning with the fourth edition in 1975, MIM went to automatic photocomposition from computer tapes and was printed in normal upper- and lower-case type. The size of the volumes increased as the amount of information grew.

MIM lists only one entry per gene locus. If there is more than one gene per locus, all those genes are listed in that one entry. This is the case, for example, with the so-called beta globin locus on chromosome 11. This is the location of the gene for sickle cell anemia and for the many beta-thalassemia genes. About 400 mutations are known for that locus. But they all are listed in one entry in MIM. Each gene locus is given a number. MIM numbers beginning with 1 are in the dominant catalog, those beginning with 2 are in the recessive catalog, and those beginning with 3 are in the X-linked catalog. The entries are arranged alphabetically in each catalog rather than by MIM number. Once a gene locus is assigned a number, it is not changed. More than 48,500 authors are associated with the diseases and conditions in MIM 8, and that goes back to 1911 and E.B. Wilson.

McKusick now wonders how he will handle this ever-increasing volume of data. The eighth edition of MIM included entries for 4,344 gene loci. In January 1989, ten months after the computer files for MIM 8 were closed, the count had risen to 4,584. That's a growth rate of 24 new gene loci per month. By the end of 1988 nearly 1,500 genes had been mapped to specific chromosomes.

The numbers might be somewhat confusing. Genes and gene loci are not the same. Loci are locations of genes, and a locus—as noted above—can contain many genes. Also, the number of genes (about 1,500) mapped to specific chromosomes is still smaller than the number of known loci. Identifying and mapping the specific location of a specific gene is more difficult than identifying a gene locus. Then there are the gene loci which contain genes associated with diseases. That number stands at around 3,000, says McKusick. However, many of these illnesses and their associated gene or genes are described only on the basis of clinical characteristics. They are known to be inherited, but no one yet knows the precise biochemical defect. According to McKusick, there are only about 400 gene loci carrying mutations for which the molecular defects have actually been identified. Finally, there are other illnesses like Huntington's disease or multiple sclerosis where researchers now have identified the genetic location, but still don't know the biochemical problem. So there are several different ways of counting up the numbers of "known" or "identified" genes, depending on how one looks at them.

SPLICING GENES

As the 1960s came to an end, one scientist decided to change the direction of his research. That decision changed the course of history. Paul Berg decided to shift his own research from prokaryotes and the viruses that infect them, to mammalian cells. Berg was a well-known researcher at Stanford University. He and others wondered if the genes and mechanisms of genetic expression were the same in higher organisms as in bacteria. The simple—or simplistic—answer seemed to be "yes," but no one knew for sure. Few researchers had spent much time actually doing experiments to answer the question, since mammalian cells are by their very nature extremely complex organisms. But Berg had a plan. He would learn about the genetic mechanisms of mammalian cells by studying the viruses that kill them. It had worked before; it might work again.

Researchers had learned a great deal about prokaryotic cells by studying the phages that infected them. Phages all tend to have a similar structure: A "head" which contains the viral DNA or RNA, and a "tail" by which the phage attaches itself to the host cell. Some phages which infect and kill prokaryotic cells had been very useful to Berg and others in their study of the prokaryotic system of genetic transmission. It turned out that some of these phages would often pick up genetic material from an infected host bacterium and carry it into another bacterium they infected. These genetic "scraps" could then become incorporated into the genome of the newly infected prokaryotic cell. Thus the phages became a powerful tool in analyzing the genetic chemistry of a living cell.

Berg and others had studied in these kinds of phages their work on prokaryotic genetics. He decided to try the same technique, studying mammalian cell genetics by looking at *their* viruses. Berg took a sabbatical from his work at Stanford and started to work on a particular mammalian virus called *SV40*. "SV" stands for "simian virus"; SV40 infects cells in monkeys and humans. It is also capable of producing tumors in certain kinds of mammalian cells. Berg hoped that mammalian viruses like SV40 might do the same thing as prokaryotic phages: pick up scraps of genetic material from one infected cell and leave it in another. However, it became clear that this was not the case. The viruses that infect mammalian cells are extremely small. They simply don't have enough room to pick up and contain a lot of DNA from one cell, and then carry it into a mammalian cell they are infecting.

At this point, in late 1970 or early 1971, still another idea occurred to Paul Berg. It was an idea that literally changed the world.

Maybe, he thought, he and his colleagues could *construct* a virus *containing genetic material from different species*. It would be a "recombinant virus," a virus whose DNA or RNA would be a recombination of different genetic material. They would put into the virus a gene or genes they might want to study. Then, if they could learn to do that in a test tube or petrie dish, *in vitro*, they might well be able to do it in mammal cells as well. They would use the recombinant viruses to incorporate the gene they wanted to study into the genome of the mammalian cell.

There seemed to be an obvious way to do this. It involved connecting some little chemical chains to the viral DNA, and then adding chemically complementary chains on the end of the piece of DNA they wanted to study. The two pieces of DNA would then naturally join together. In this way one could construct a recombinant DNA molecule. It seemed obvious, and easy.

And it was.

Berg and his laboratory colleagues set out to do it. It was done within eight months. Berg thought it was a "neat experiment." He didn't think it was earth-shattering, but rather fairly straightforward and obvious. He also thought it would eventually open the way to construct recombinant DNA molecules and viruses with any other kind of genes that might become available. Of course, in 1971 there *weren't* other genes available. Gene cloning had not yet been invented.

Today, Berg feels that the experiment was not one that could have been done in too many other places. First of all, his laboratory at Stanford was one of only a few that had all the chemicals and enzymes needed to do the experiment. Also, the crux of the problem was the idea of molecular "sticky ends." And that was something which few researchers other than Berg knew about. In the 1960s Har Gobind Khorana had discovered that an enzyme made by a bacterial virus called T4 could link DNA molecules together. The T4 enzyme essentially acted like a kind of chemical glue. Khorana had used the T4 enzyme to stick together pieces of DNA, but they were DNA fragments from the same species.

Berg's research team consisted of himself and one post-doctoral student, David Jackson. Jackson had already spent a year working on another project when Berg convinced him to try this experiment. Jackson, in fact, was the one who did the work. With a little help from an Australian researcher named Robert Symonds, he went one step beyond Khorana's work. They first randomly chopped up the DNA from the SV40 mammalian virus and from a bacterial virus called lambda. They didn't know where in the viral DNA the pieces would fit and function. So they wanted to cut open the DNA in random places and see what happened. Next, they used various enzymes to make "sticky ends" on the DNA segments. Then the team used the T4 enzyme to glue the two segments together. The result was the world's first recombinant DNA molecule—a hybrid DNA molecule made from pieces from two different species of living creature.

This was not the only way to create recombinant DNA molecules, of course. Different *restriction enzymes*, chemicals that "restricted" the length of a DNA molecule by cutting it at different base-pair locations, had been discovered before Berg and Jackson started their gene-splicing experiment. One such restriction enzyme is called *Eco RI*. (The "Eco" meant the enzyme came from the bacterium *E. coli*. "RI" meant restriction enzyme Roman numeral I.) It had been discovered in Herbert Boyer's lab at the University of California at San Francisco.

The report on Berg and Jackson's experiment was published in the *Proceedings of the National Academy of Sciences*. The journal had made an exception and published the paper even though it was longer than their maximum length. Berg knew it would create a stir. But he merely expected other researchers to recognize their work as a nice, neat experiment. He never foresaw the blossoming of a multi-billion-dollar industry within fifteen years, and a Nobel Prize for himself within ten. All Paul Berg had really wanted to do was find a way to introduce genes into mammalian cells so he could study how those genes operated, how they "were expressed" in genetic jargon.

The key to the recombinant DNA technology of today, however, is not merely splicing different genes together, but the techniques of *cloning* genes, of getting cloned recombinant genes to produce their particular proteins, and of selecting those specific cells that contain the cloned and expressing genes. Without these procedures, gene splicing is not much more than an elegant biochemical trick. The technology for cloning genes—making many identical copies of a particular gene or gene fragment—was developed in the early 1970s. The inventors were Herbert Boyer and Robert Heming at the University of California, San Francisco, and Stanley Cohen and Annie Chang at Stanford. These researchers exploited *plasmids*, circular pieces of DNA. They used them like molecular trucks to carry genes of their choice into target cells. The cells took the plasmid and incorporated its genes—including the "test gene"—into their own genome. The "test gene" would then be expressed—that is, it would create the protein it coded—by the cells as they grew and multiplied. In particular, the four researchers created plasmids carrying genes to make the target bacteria resistant to the antibiotic tetracycline. The bacteria that actually incorporated that gene into their own DNA, and expressed it, were therefore easy to identify. They were the ones not killed by doses of tetracycline.

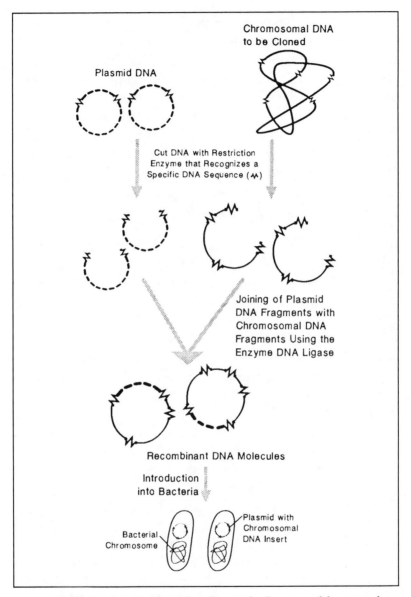

DNA Cloning in Plasmids. This method was used by several researchers, including Paul Berg, to create the first genetically engineered organisms.—Reprinted from: U.S. Congress, Office of Technology Assessment, "Mapping Our Genes—The Genome Project: How Big, How Fast?" OTA-BA-373 (Washington, D.C.: U.S. Government Printing Office, April 1988.)

The ultimate point of Berg's experiments was to take the hybrid molecule created by splicing SV40 virus and lambda virus DNAs together, and introduce it into *E. coli*. This could take place because the lambda virus was capable of infecting *E. coli* bacteria. As the infected *E. coli* reproduced, it would make many copies of the recombinant DNA molecule inside it. Berg would then have a continuing supply of SV40-lambda hybrid DNA molecules to use in further experiments on gene expression in mammalian cells.

CONCERNS, AND A "MORATORIUM"

In the summer of 1971 Janet Mertz, one of Berg's graduate students, attended a scientific conference at Cold Spring Harbor, New York. She described Berg's plan to other scientists there. The response was immediate—and negative. SV40 causes cancerous tumors in human cells, at least in the test tube, several researchers pointed out. And *E. coli* live in the human intestine. If infected *E. coli* should escape from the laboratory, and enter into humans, the results could—*could*, it should be emphasized, since no one really *knew*—be catastrophic.

Berg listened to the arguments against his planned experiments, and he agreed with them. He decided not to go ahead with the experiments as planned.

The matter might have rested there, if the science of recombinant DNA had not advanced. But it did. In 1972 Mertz, Ronald Davis, and Vittorio Sgaramella, all at Stanford, made a breakthrough with the Eco RI restriction enzyme. The two research teams discovered independently that DNA with "sticky ends" was automatically produced when the DNA *was cut with Eco RI*. The new splicing method was simplicity itself. All one had to do was put together two DNA pieces that had been cut by the same restriction enzyme, and they would stick together. For Berg's work, the task was made even easier by the fact that Eco RI cuts the SV40 viral DNA in one—and only one—location. (Imagine you have a long, thin piece of paper taped together into a loop. That's the DNA molecule of the SV40 virus. Now, imagine you have a pair of scissors. An ordinary pair of scissors can cut the paper loop anywhere. But this is a very special kind of scissors. It will cut the paper loop at only one place. This "magical" pair of scissors is the Eco RI restriction enzyme.)

Berg and his colleagues cut open the SV40 DNA loop with Eco RI. They also cut open the lambda virus's DNA (which is also circular) with the same restriction enzyme. Then they joined the two together into a hybrid creature. And the DNA segments expressed themselves—they proceeded to produce their specific proteins.

The issue of whether mammalian tumor viruses like SV40 were potentially pathogenic in humans surfaced again, and seriously. In response, Berg, James Watson, and several other researchers convened a small meeting to discuss the potential danger of working with mammalian tumor virus. Their consensus was that with modest precautions in the lab, such experiments would pose no danger to either the researchers or the public at large.

Then came the next advance in knowledge. John Morrow, one of Berg's graduate students who was working with Herb Boyer and Stan Cohen, showed that one could introduce a *totally foreign gene* into E. coli and that the gene could function. Morrow put a frog gene into a bacterial plasmid and demonstrated that the frog gene would be expressed. At that point, it began to look like anybody could put in any kind of DNA into anything, and the gene would produce its protein product. Suppose it were in fact that simple, many scientists began to worry. Suppose, say, the genes for the botulin toxin could be spliced into the ubiquitous E. coli, and the E. coli released into the general population? These kinds of scenarios led the eminent biologist Maxine Singer and others, at the June 1973 Gordon Conference in New Hampshire, to express their concern to the National Academy of Sciences, and its president, Philip Handler.

Handler's response was to ask Berg to convene a study group and examine the potential benefits and risks of recombinant DNA technology, and make some recommendations on how he and the Academy should proceed. Berg went to work with James Watson and several other eminent scientists. In 1974, after a year of study and debate, the group released their conclusions in the form of a letter published in *Science* magazine. In essence, the letter (now often referred to as "the Berg Letter") said that great opportunities existed with recombinant DNA technology. But there were also many unknowns, and perhaps some risks. Since we don't know for sure if there *are* any risks, we ought to take a very close look at this new biological technology before we proceed any further. So let's get a bunch of experts together and figure out how to proceed. The letter also called for a "deferral" of certain kinds of experiments until the matter was settled:

■ No experiments that would put genes coding for toxins into bacteria infectious to humans.

■ No experiments that would put genes coding for antibiotic resistance (like the Boyer-Cohen experiment) into organisms infectious to humans, and which might therefore interfere with the use of those antibiotics.

■ No experiments that would knowingly put oncogenes—cancer-causing genes—into these organisms.

All other recombinant DNA experiments, the letter suggested, would be fine.

The popular news media have frequently referred to the request for a deferral of certain experiments as a "moratorium." However, Berg himself has always disliked the use of that word. The scientists in the study group spoke out not because they were certain there *were* risks. There was never a sense that any researcher *knew* there was a hazard from recombinant DNA experiments. No hazard had actually been seen, nothing terrible had happened. Rather, says Berg today, they called for a deferral out of their ignorance. They didn't know if there were risks in these kinds of experiments or not. The bottom line, says Berg, was that no one knew for sure one way or another. Utter uncertainty and lack of knowledge led to the call for caution. That call by Berg and Watson was not a ban or a moratorium on all recombinant DNA work, but only on three kinds of experiments that seemed to have the greatest potential risks.

It is tempting to compare the Berg Letter to another letter from a famous scientist, the one written by Albert Einstein to President Roosevelt in the late 1930s urging the president to consider the development of an atomic bomb. The similarities, however, are mostly superficial. Einstein was already a Nobelist and world-renowned figure when he wrote his letter. Berg was well-known in his field, but not outside it, and didn't win his Nobel until nearly a decade later. Einstein was motivated by a fear that the Nazis would develop and use the atom bomb before anyone else. Berg was concerned not about war or weapons, but about scientists carrying out scientific experiments in a knowledge vacuum. Both atomic energy and genetic engineering have enormous potential for both good and evil. However, Einstein knew full well that the decision to create an atomic bomb, a weapon of mass destruction, was at best morally ambiguous. Berg's letter was not intended to *encourage* the development of a technology intended for destruction, but to *delay* the development of a technology on the grounds of ignorance.

The result of the Berg Letter was the convening of the international gathering today known as the Asilomar Conference. It took place in February 1975 at the Asilomar Conference Center in Pacific Grove, California.

Most of the Asilomar conference was purely scientific in nature: researchers reporting on their work to other researchers. It was only on the last morning that the assembled scientists launched into a serious discussion of the potential risks and how to deal with them. The organizers of the conference had spent the previous night working on a series of research guidelines and recommendations. The general feeling was "We still don't know enough about this field of endeavor, so we better proceed cautiously." There was also a feeling that scientists needed to be educated about what the potential risks might be in recombinant work. Then, as they proceeded, learning which risks are real and which are imagined, they could change the guidelines to fit the realities of recombinant DNA research and its known risks.

The recommendations, then, were that the "moratorium" be lifted and replaced with a set of special safety procedures for scientists to follow. The procedures followed would depend on the kind of DNA experiments being done. Not everyone agreed with this proposal. But the great majority of those in attendance did agree with the recommendations. The report was submitted to the National Academy of Sciences and published in *Science* in July 1975. It became the basis of the genetic engineering guidelines later promulgated by the National Institutes of Health, which have governed recombinant DNA research since 1976.

Over the years the guidelines have been successively modified and relaxed. It has never been necessary to impose more stringent guidelines. Since the first gene-splicing by Paul Berg in 1970 and 1971, and after millions of experiments done around the world, not the slightest scintilla of evidence of risk to anybody has surfaced.

One of the fears expressed by some opposed to guidelines and restrictions on gene-splicing was that they would harm the scientific process or impede "scientific progress." In retrospect, Berg and others are convinced that science neither suffered damage, nor was impeded. In fact, Berg thinks that science and scientists learned a lot from the Asilomar process, and were able to convince the public at that time that they were not trying to endanger the planet in order to satisfy their own whims.

SANGER AND GILBERT SPEED SEQUENCING

When Paul Berg won the 1980 Nobel Prize for Chemistry, he split the award money with two other researchers, Frederick Sanger and Walter Gilbert. Sanger and Gilbert received their share of the Nobel for developing new and quicker ways of sequencing DNA. "Sequencing" is the process of determining the order of nucleotides—of As, Ts, Cs and Gs, the letters of the genetic alphabet—along a DNA strand. Gilbert and Sanger, in other words, figured out ways to read DNA more quickly than those previously available.

Frederick Sanger, of Cambridge University in England, had won a Nobel Prize in 1958 for figuring out how to sequence proteins. Then, in the 1960s, he developed a new and accurate method of sequencing RNA. At the time RNA was easier to sequence than DNA. RNA is shorter, and sequencing could be done on single stranded RNA. Also, as noted earlier, it is RNA which actually carries the coded protein sequences to the cell machinery which creates proteins. Because of Sanger's earlier work, researchers around the world were able to read the code from more RNA sequences than possible before. Reading genetic codes was becoming easier, cheaper, and quicker. However, DNA was still too difficult to sequence in any significant length. In the early 1970s Sanger took the next step. He set out to develop a DNA sequencing process.

His method involved synthesizing segments of DNA one nucleotide at a time. One begins with a cloned fragment of the DNA whose sequence is unknown. It is mixed with a short piece of synthetic DNA which will match with only one end, the origin end, of the fragment. The enzyme called DNA polymerase is then added to start the synthesis of a complementary strand of DNA. This takes place in a mixture of nucleotides (A, T, C, and G), one of which is tagged with a radioactive phosphorus or sulfur atom. The mixture also includes a modified nucleotide called *dideoxynucleotide*. The modified nucleotide (again either A, T, C or G) stops the creation of the complementary strand whenever it is randomly inserted. Four

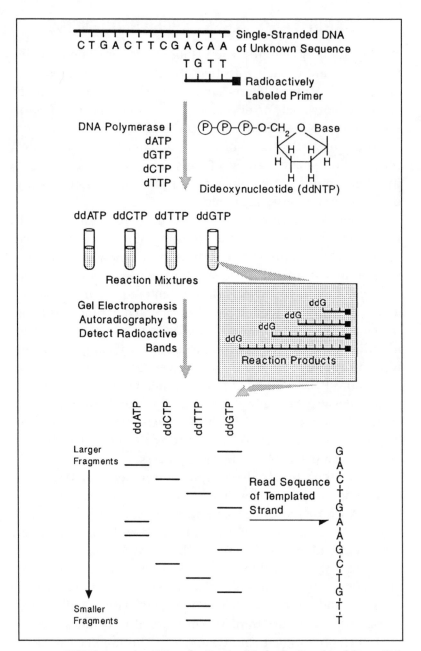

DNA Sequencing Using the Sanger Method.—Reprinted from: U.S. Congress, Office of Technology Assessment, "Mapping Our Genes— The Genome Project: How Big, How Fast?" OTA-BA-373 (Washington, D.C.: U.S. Government Printing Office, April 1988.)

separate reactions can be run, each using a different modified nucleotide. The result will be the creation of a series of radioactively labeled DNA segments of the "mystery fragment," each ending at a different nucleotide. The series is like a set of "nested" segments, pieces of the entire DNA strand which have overlapping sequences. The segments can be separated with the standard technique called gel electrophoresis. The sequencing reactions for each of the four bases are run as adjacent lanes on the gel. A piece of X-ray film is then placed over the gel block, and the fragments photograph themselves because of the radioactive atoms attached as labels. The *autoradiograph* thus created shows a ladder-like pattern of bands. The DNA sequence is deduced from the patterns in the lanes revealed on the autoradiograph. This is a time-consuming process using sometimes-dangerous radioactive isotopes as tags. But at the time it was a great leap forward in sequencing DNA. Sanger would later go on to be the first person to sequence the genome of a virus. In 1977 he sequenced the phage called ϕX174 ("ϕ" is the Greek letter "phi").

Where Sanger had consciously set out to find a way to sequence DNA, Walter Gilbert, of Harvard University, did not. In fact, it happened almost by chance. He had previously done research on a protein made by *E. coli* which interacted with a particular segment of DNA and turned it off. The segment was an *operon*, a series of genes that work together to create different proteins with related functions. The particular operon Gilbert was interested in is called *lac*. The protein is the *lac repressor protein*. At the urging of a Russian researcher, he and his colleague Allan Maxam became interested in new ways of studying how these kinds of proteins recognize the operons with which they are supposed to interact. They were using a chemical called dimethyl sulfate, which cuts DNA sequences at the nucleotides adenine and guanine.

The DNA sequence of the lac operon was already known, since the genes had been copied to RNA, and the RNA version had been sequenced. However, no one knew exactly where the lac repressor protein attached to the DNA. The nucleotide sequence of that part of the DNA was not known. Maxam and Gilbert decided to find out. They first exposed DNA containing the lac operon to the chemical and broke the molecule into pieces ending with As and Gs. As a comparison, they then took some more lac operon DNA, exposed it to the lac repressor protein so the protein would bind to the lac operon, and added dimethyl sulfate. The As and Gs in the DNA which had reacted to the repressor protein would not be affected by the chemical, and the DNA did *not* break apart at those particular locations. They were protected by the protein. Gilbert and Maxam could find out where the repressor protein connected to the DNA by finding the DNA segments that didn't end with As or Gs, but had them inside.

Later, Gilbert and several other colleagues discovered another lac operon in the *E. coli* DNA. They did not know its sequence. Maxam repeated the procedure with dimethyl sulfide and lac repressor protein, finding the area where the protein bound to that area of DNA and thus the location of the new operon.

And at that point, he and Gilbert realized they had stumbled onto a new way to sequence stretches of DNA. By fiddling a bit with their technique, they could adjust it to cut DNA at either adenine or guanine. After working through the summer of 1975, Maxam eventually found a chemical that could, with appropriate adjustments, cut DNA at either the thymine or cytosine. The Maxam-Gilbert chemical method of sequencing DNA was born.

The Maxam-Gilbert and Sanger DNA sequencing methods revolutionized genetics. The two sequencing techniques made it possible for researchers to decode segments of DNA much more quickly than ever before. The Maxam-Gilbert and Sanger DNA sequencing techniques, the cloning methods developed by Herb Boyer and Stan Cohen, and the essential technology

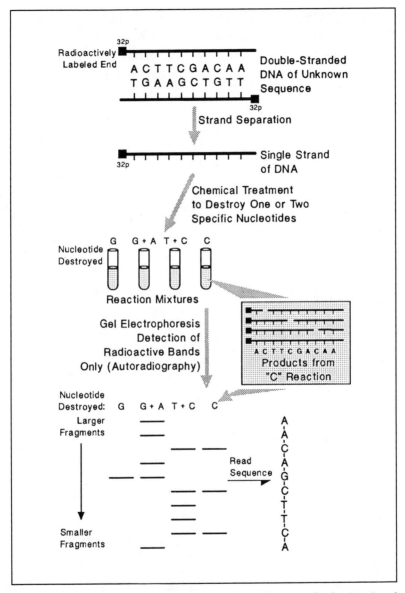

DNA Sequencing Using the Maxam-Gilbert Method.—Reprinted from: U.S. Congress, Office of Technology Assessment, "Mapping Our Genes—The Genome Project: How Big, How Fast?" OTA-BA-373 (Washington, D.C.: U.S. Government Printing Office, April 1988.)

of gene splicing initiated by Paul Berg all led directly to the genetic engineering revolution of the 1980s.

And without these singular developments of the 1970s, the Genome Project would not be possible.

4

Momentum

"$3,000,000,000"

WALTER GILBERT

T ODAY, DNA can be sequenced with considerable accuracy. The sequencing techniques of Gilbert, Maxam, and Sanger have made it possible for scientists to identify many different functional regions of DNA, as well as areas of DNA which seem to be "silent"—so–called "junk DNA."

And they have led to a huge increase in the production of DNA sequences. In the 1970s, George Bell of the Los Alamos National Laboratory (LANL) put together the Lab's Theoretical Biology and Biophysics Group. That group, along with ten others, is now part of LANL's Theoretical Division. In that Division is located a bank of computers which electronically store different DNA sequences discovered by researchers around the world. It is called the GenBank. In recent years it has been swamped by the influx of sequences. This is despite the fact that DNA sequencing is a laborious process, done by hand and often with the use of dangerous radioactive chemicals as tracers attached to DNA segments.

One major reason for the vast increase in sequenced DNA segments was the discovery in the late 1970s of a way to create genetic maps using special markers or mileposts within chromosomes. And one of the movers and shakers of this technique was Dr. Raymond White.

MARKERS, RFLPs, AND RAY WHITE

Ray White is a soft-spoken handsome man who heads the Howard Hughes Medical Center at the University of Utah in Salt Lake City. He is one of the major players in the Genome Project. White and his colleagues, with the enthusiastic participation of several dozen multigenerational Mormon families, are sketching out some of the first detailed maps of parts of the human genome. White is doing it with a technique he helped pioneer for use in human genetic mapping, the use of *genetic markers.*

White completed his undergraduate studies in molecular biology at the University of Oregon in Eugene in the early 1960s, at a time when the university's Institute of Molecular Biology was a new and exciting enterprise. After graduation he was urged to go to MIT and work with Morey Fox, a decision that shaped White's scientific career. He learned recombinant DNA techniques under Fox, working with lambda phages, just as that revolutionary genetic technology was being invented by people like Paul Berg, Herbert Boyer, and Stanley Cohen. The result was that Ray White, in the mid-1970s, was one of a handful of people possessing a solid background in the biochemistry of recombinant DNA technology. It was altogether appropriate that he would pursue laboratory tests of the hypothesis that these genetic markers might exist, and that they could be used to create a map of the human genome.

White first heard of the idea of making DNA markers for the human genome in the spring of 1978 through his thesis mentor, Morey Fox, who is now the chairman of the Biology Department at the Massachusetts Institute of Technology (MIT). Fox had just returned from a breast cancer task force meeting in Washington, D.C. That meeting was also attended by Mark Skolnick from University of Utah. Skolnick in turn had come to the meeting right after attending a scientific retreat of the University of Utah's genetics program people. Also there were David Botstein of MIT and Ronald Davis of the University of California. Together these four would pioneer the use of genetic markers to map the human genome.

The idea of finding genetic markers, something akin to mileposts on a highway, is based on the action of enzymes, chemical compounds that causes changes in other compounds, such as proteins and peptides. There are many different kinds of enzymes, and millions of specific ones. DNA restriction enzymes, like the Eco RI enzyme discovered by Herbert Boyer,

Ray White—articulate, sometimes controversial, and a leader in the effort to create genetic linkage maps of several human chromosomes. A scientific paper coauthored by White in 1978 set the stage for the development of the Genome Project.—Photo by Joel Davis.

David Botstein was instrumental in writing, with Ray White, the 1978 paper that suggested using RFLP markers for genomic mapping. He was later involved in the work by Collaborative Research, Inc. in genetic linkage mapping of the human genome. Botstein went on to become a vice president at Genentech, Inc., one of the world's top genetic engineering companies.

chemically recognize a particular DNA sequence—that is, a specific arrangement of A, T, G, and C bases which make up the DNA alphabet. They then cut the DNA strand at that particular point. Each DNA restriction enzyme cuts the DNA at a different location. DNA restriction enzymes can thus be used as chemical "knives" that have the ability to cut DNA into fragments of specific and predetermined lengths and sequences. It is as if a pocket knife would cut a piece of salami only into three-inch fragments, while a butcher knife would cut the same salami only into pieces seven inches long. The fragments of DNA created with restriction enzymes are called *restriction fragments*. Researchers can separate the restriction fragments by size using the standard technique of gel electrophoresis.

Restriction fragments created with the same enzyme will differ in length from person to person because each person has a unique set of genes. DNA sequences for the same gene can differ from person to person. Sometimes just one base pair is different; sometimes a gene in one person has an extra one or two base pairs; sometimes an entire block of DNA may be missing, or an extra block of DNA may be present. Whatever the reason, the result is the same. The cutting site for the restriction enzyme is changed or lost, or a new one is created. The enzyme cuts the DNA at a different place, and a different-sized restriction fragment results. Molecular biologists call these kinds of genetic differences *polymorphisms*, from the Greek roots "poly" ("many") and "morph" ("shape"). The polymorphisms cause changes in the lengths of the restriction fragments. Such a DNA fragment is called a *restriction fragment length polymorphism*, or RFLP, pronounced "riflop."

Different RFLPs caused by genetic variations can all be identified and separated out with the gel electrophoresis process. Another identification tool uses pieces of DNA called probes which have been marked with short-term radioactive tracer elements. Researchers can use these radioactive probes to identify and isolate RFLPS with *specific genetic sequences* from a large collection of DNA fragments. This is not a new technology. The method was first developed in 1975, and by 1977 was being used to create a map of DNA regions near a particular gene in the rabbit genome. This technology, nearly as old as genetic engineering itself, has made it possible to identify variations from person to person within a specific part of the human genome. In other words, it is possible to tell the difference in a part of the genome between a person who has, say, a genetically-caused disease like Huntington's disease, and a person who does not.

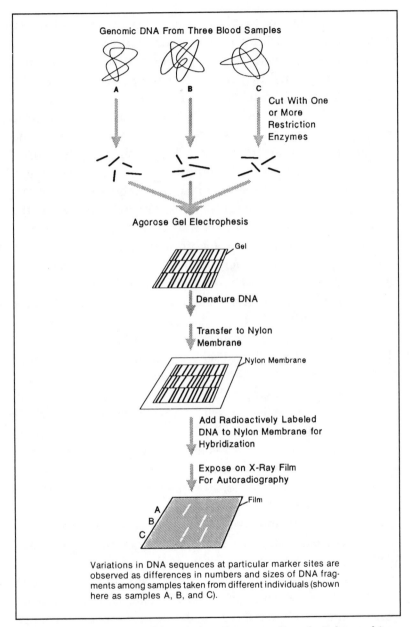

Genomic DNA From Three Blood Samples

A   B   C

Cut With One
or More
Restriction
Enzymes

Agorose Gel Electrophesis

Gel

Denature DNA

Transfer to Nylon
Membrane

Nylon Membrane

Add Radioactively Labeled
DNA to Nylon Membrane for
Hybridization

Expose on X-Ray Film
For Autoradiography

Film

A
B
C

Variations in DNA sequences at particular marker sites are
observed as differences in numbers and sizes of DNA fragments among samples taken from different individuals (shown
here as samples A, B, and C).

The Detection of Restriction Fragment Length Polymorphisms
(RFLPs) Using Radioactively Labeled DNA Probes.—Reprinted from:
U.S. Congress, Office of Technology Assessment, "Mapping Our
Genes—The Genome Project: How Big, How Fast?" OTA-BA-373
(Washington, D.C.: U.S. Government Printing Office, April 1988.)

At the Utah retreat, Skolnick and his people had talked about their work with a human genetic disease called hemochromatosis. (Some research papers on hemochromatosis had also been published by Jean-Marc Lalouel, who would later work with Ray White at Utah. At that time, however, Skolnick was unaware of Lalouel's work.) Sometimes called "bronze diabetes" because of the skin coloration which occurs, hemochromatosis is a disorder in which too much dietary iron is absorbed by the body. It is confined almost entirely to men. Women are rarely affected because they regularly lose excess iron during their menstrual periods. For men, however, the excess iron slowly accumulates in their heart, liver, pancreas, testes, and other organs. Hemochromatosis rarely causes problems until a man has reached middle age. First sexual drive decreases and the testes begin to shrink. Then the excess iron causes cirrhosis of the liver. The pancreas is damaged and is unable to produce enough insulin, leading to diabetes mellitis. The bronze skin color is caused by the buildup of iron in the skin. The late stages of the disease are characterized by serious cardiac problems, liver failure, and liver cancer.

Researchers studying hemochromatosis had had a difficult time figuring out what caused it. Though a few researchers doubted that it *was* genetically caused, that was the general scientific consensus. But—was it a frequent recessive allele, a rare dominant allele, or what? And more immediately—exactly where was the gene, anyway? Skolnick and his colleagues had used a genetic marker system to track down the genes for hemochromatosis. The markers they used were ones that already exist: the ones for the HLA immune system complex of genes. These genes are located on short (or "p") arm of human chromosome 6, and proved invaluable in understanding the inheritance of hemochromatosis. First, the HLA genes contain a huge number of alleles, or variations in genes, from person to person. That's what helps make each person's immune system unique to that person. Those allelic variations are inherited. Furthermore, the hemochromatosis gene was suspected to be located fairly close to the HLA complex. By studying the inheritance pattern of HLA alleles in different families, Skolnick was able to show that hemochromatosis was a genetically recessive disease. In fact, the evidence now is that the gene for hemochromatosis actually sits somewhere *between* two parts of the HLA complex of immune system genes, HLA-A and HLA-B.

THE RFLP PAPER

During a conversation at the Utah retreat with Davis and Botstein about his hemochromatosis work, Skolnick was bemoaning the fact that there's only one HLA complex in the human genome. What he meant was, wouldn't it be nice if we had more sets of genes like HLA, with lots of variations in the genes, located on many other chromosomes besides number 21. Furthermore, wouldn't it be wonderful if all those variant genes and gene clusters were well known and well-mapped. Then researchers could use them to follow the inheritance patterns of other genetic diseases. At that point Botstein and Davis essentially said: "Hey, wait a minute. There's no reason to wish for or look for a lot of HLA markers in the genome. You can make your measurements by using restriction enzymes." There then followed a day or so of intense discussion among Botstein, Davis, and Skolnick of different ways to approach that kind of project and do it in one fell swoop. None of those ideas immediately matured into a real project. But at that point an important question was being asked: How can we do this?

Skolnick mentioned all this to Fox at the Washington, D.C. meeting. Fox then told Ray White about the mapping idea. White was appropriately trained in the technologies needed to measure the location of gene markers using restriction enzymes. When he heard about the concept, he was fascinated. White at that time was looking for something to do, for a new challenge. He had long wanted to get involved in human molecular genetics, and had been waiting for some window of opportunity to open up for him. Now, with Fox's description of Skolnick's conversation about mapping gene locations using restriction enzymes, the casements swung open. White jumped through, and into a career that has led him to the creation of detailed maps of different human chromosomes.

White and Botstein knew each other well. Botstein had begun his thesis work with Fox, although he later finished it at the University of Michigan, and then returned to MIT to teach. White and Botstein had worked in adjacent labs during White's later graduate student years. The day after hearing Fox's story, White called him. The timing couldn't have been better. Botstein was already thinking about ways to promote the nascent project of genetic mapping with genetic markers. And promoting research projects is something he had always been good at, and still is.

White recalled Botstein's response to his call as "quintessential Botstein."

"Wonderful," Botstein supposedly said to White. "I'm so glad you called. You're the *second* person I had on my list!" The comment wasn't meant to be mean or vicious. It was the kind of "off-the-cuff" candor often typical of Botstein. In fact, White felt quite honored to be told he was "the second person" on Botstein's list—and he still doesn't know who the first person was. Of the several people excited by the idea of this kind of project, he was the one with the experience and willingness to commit a laboratory to do it. At that point he wrote an NIH grant to fund the project. Botstein gave him several comments, and signed as a co-investigator.

The four people who wrote the RFLP paper were uniquely suited to the task ahead. David Botstein is a genius at pulling together a grand vision, and Ron Davis has a deep scientifically intuitive imagination. Mark Skolnick is a man who bridges many disciplines and speaks many scientific languages. White has described himself as "pretty good" at intuiting nucleic acid technology.

At first, the four men saw the idea as a straightforward scientific endeavor, and talked about it with their molecular biologist friends. They got plenty of comments that fell into the category of "It'll never work, and here's why," with what White later called "a bushel basket of reasons." That is a fairly typical response to a project that is only starting to get off the ground. In fact, it's a good response for scientists and science. If a scientific theory cannot stand up to criticism, then it is not a good one. Criticism is one of the ways science works.

It was Skolnick and Botstein who started pushing the idea of writing a theoretical paper about genetic mapping with RFLPs. Ray White did not feel there was much theory to write about. This was not like discovering a new class of subatomic particles. It was just as revolutionary, however. White and his colleagues were essentially telling the scientific world: "If we do the following, we ought to be able to make genetic markers and a genetic map, and then map the location of human genetic diseases." The authors then proposed that this work be done, that the markers be discovered and used to make a such a genetic map. The paper they ultimately wrote didn't have an immediate impact. In part that was because the journal in which it appeared, *The American Journal of Human Genetics*, is read mainly by human geneticists, a small subset of the genetic research field. The journal's readers were impressed by it, but didn't understand the technology, and in any case could do little about it. For the first year after its publication, then, the Botstein-Davis-Skolnick-White paper received a restrained reception. Nevertheless, the authors themselves believed it was truly important from a medical and genetic point of view.

The four researchers realized others in biology were starting to feel the same way in 1979. White presented some of their ideas and early experiments at a meeting that year at the Cold Spring Harbor Laboratory, now headed by James Watson. After that meeting, some of the molecular biologists complained to White that "that report of yours is all so obvious." It was at that point, when his colleagues' irritability was beginning to surface, that White realized they were catching on to the importance of their idea. Researchers rarely complain about new ideas they consider *unimportant*.

Just how important their proposal was became clear six years later.

THE SANTA CRUZ MEETING

In May 1985 a group of scientists gathered at the University of California at Santa Cruz to toss around the idea of mapping the human genome. Most of them thought the idea was crazy. When the meeting was over, most of them still thought it was crazy. But they also thought it could be done, and within their lifetimes. It was a turning point for the Genome Project.

The Santa Cruz meeting was organized by Dr. Robert Sinsheimer, then the chancellor at University of California at Santa Cruz and now a professor of molecular biology at University of California at Santa Barbara. Sinsheimer had spent twenty years in the Biology Division of the California Institute of Technology, nine as its chairman. (The next person to be chairman of that division was Leroy Hood, another major figure in the Genome Project). Ten years earlier he had been one of the voices urging researchers to desist from the rush to splice genes. He had attended the Asilomar Conference, worried by concerns about the potential hazards of gene splicing. The intervening years had brought new knowledge about genetics, and Sinsheimer's concerns had been laid to rest. Meanwhile, in the early 1970s Sinsheimer's own lab had purified and created a genetic map of the ϕX174 phage, making it possible for Fred Sanger to sequence its genome of 5,400 base pairs in 1977. By 1985 scientists had sequenced the genomes of bacterial viruses nearly ten times larger, such as T7 (40,000 base pairs) and lambda (49,000 base pairs). The sequencing of the Varicella-Zoster virus (125,000 base pairs) was far along. Researchers were beginning the long and tedious effort to sequence the genome of *E. coli* (4.5 million base pairs) and even of the nematode (80 million base pairs). Sinsheimer was well aware of the hardships in such sequencing projects—and the trends.

Now he was proposing the use of recombinant DNA technology to map the genome. His reasons were simple. Sinsheimer had never been greatly enamoured of the "big science" of other disciplines like astronomy and physics. But he wondered if biologists might not be missing some important opportunities because they were not thinking in terms of large-scale projects.

What's more, Sinsheimer had experience in organizing a "big science" project. While chancellor at University of California at Santa Cruz, he was involved in the preliminary efforts to snag the Superconducting Supercollider (SSC) for the state of California. The SSC will be the next generation of particle accelerator for high-energy physics. By smashing beams of protons and antiprotons together at trillions of electron-volts of energy, physicists hope to peer into the ultimate heart of matter.

Another project Sinsheimer was involved in, and with considerable hands-on activity, was astronomical in scope. It was a proposal to build a 10-meter telescope. The telescope would have a mirror more twice as large as the five-meter telescope at Mount Palomar in Southern California. It would use the latest in computer technology and would be one of the largest in the world. Sinsheimer worked hard to convince a multimillionaire to make the monetary commitment to the project, and he helped organize the necessary political backing.

In the end, the initial financial proposal for the telescope project fell through, in large part because of various bureaucratic problems. The telescope, now called the Keck Telescope, did eventually enter construction as a joint project of Caltech and the University of California, with funds from the Keck Foundation. Meanwhile, Sinsheimer, a molecular biologist, became intrigued by the idea of mapping the location of all the genes in the human genome. Perhaps the way to do it was to take the Big Science route. The first thing to do, however, was to determine if that was feasible. One way of doing this is to get a bunch of smart people together to talk about it. Which is just what Sinsheimer did.

In the fall of 1984 the money for the super-telescope, some $36 million, had been returned to the donor. Sinsheimer, who for several years had been thinking hard about the potential application of genetic knowledge to the human condition, then hit upon an idea. Perhaps the $36 million could be used to launch an "Institute to Sequence the Human Genome" at University of California at Santa Cruz. He suggested this in a letter to the President of the University of California. It was turned down. Sinsheimer felt the basic idea was still a good one. To get other funds, though, he would need a rational plan of attack, a concept of the pace and scale of such a project in terms of both time and money. He needed some way to validate his concept of a "genome project."

Sinsheimer met with three molecular biologists at UCSC: Harry Noller, Bob Edgar, and Bob Ludwig. They discussed setting up a workshop on the feasibility of sequencing the genome, outlined a strategy for carrying it out, and drew up a list of names of people to invite.

On Friday evening, May 24, 1985, about a dozen biologists showed up at the California town to talk about genetics and Big Science. They included Leroy Hood of Caltech, whose lab had just invented four machines to sequence or synthesize segments of DNA or proteins; John Salston of the National Research Council; Michael Waterton of the University of Southern California; and David Botstein and Walter Gilbert. Several people from University of California at Santa Cruz also took part. They settled into the Fellows room of Crown College and talked, argued, laughed, puzzled, and debated. The two-day meeting touched on every issue and argument that would surface in subsequent meetings and debates: Is it feasible? Could such a project even be started? What about funding? Is a "Big Science" approach a wise one for sequencing the genome? Should we sequence the "junk DNA" in the genome?

To the last two questions, Sinsheimer's answers were "yes" and "yes." While there is nothing virtuous in itself about "Big Science" projects, in this case it could create for biology something it desperately needed: a massive information base of knowledge about the genetic structure of key organisms. As for "junk DNA," Sinsheimer's feeling was that "one person's junk is another person's treasure." Analyzing the sequence of such regions would be valuable for evolutionary and anthropological research. Also, coding regions, control regions, and "junk" regions are all mixed up together in the human genome. Only by sequencing it all could biologists really determine what was "junk" and what was "garbage."

The workshop attendees finished Sunday morning, May 26, convinced that a human genome project was feasible. There was plenty of work to do, much technology to be invented, and many people to be convinced—especially government people with funding. But it was feasible. Sinsheimer later wrote up a summary of the workshop and began distributing it. His Santa Cruz meeting was the first significant gathering of researchers in pursuit of the Genome Project. It would not be the last.

RFLPs REDUX

That same year, 1985, Ray White and several others wrote a paper which appeared in *Nature*, on the subject of RFLPs and genetic mapping. This paper presented the substantial progress that had been made on several fronts since White and his colleagues had first proposed the idea of RFLP genetic maps in 1978. The new report included a description of the families White had used to collect genetic data, as well as a discussion of what kinds of families would be the right kind for a genetic mapping project using RFLPs.

White and his co-authors had first thought that the best families to use would have extended pedigrees—that is, with individuals from many living generations. These kinds of families are indeed the best when one wants to gather data on *genetic disease linkages*. They are useful in large part because one can be sure that one gene is actually causing the disease in that family. However, that was not the specific goal of White's project. He simply wanted information that would lead to a genetic linkage map. It turns out that families with a well-defined *three-generation* structure are very useful for this kind of information. This is especially true when each of the families has a lot of children. Not only is it efficient—researchers would get a lot of information in terms of the amount of work they did—but working with a three-generation family is also easier in terms of the calculations the researchers have to perform.

Another important aspect of White's 1985 paper was the discussion of the significance of *multilocus analysis*. White and French genetic research Jean-Marc Lalouel both realized the importance of being able to look at the inheritance of *several genetic loci* at a time. This kind of analysis is conventional in the genetic study of other species, but at that time was not often used in human genetic studies. There weren't a lot of genetic markers to make multilocus analysis feasible in humans, for one thing. For another, the analytical power was not yet available. The computer and mathematical programs used to study genetic linkages in humans were not then adaptable to simultaneous studies of several human genetic loci.

By the early and mid 1980s, though, it had become critical to know the order of genetic markers in a chromosome. Multilocus analysis would make this possible, and White's 1985 paper made this point. Indeed, White himself had already written some small computer programs so he could analyze some of his data. Unknown to White, Lalouel in Paris was already beginning to use a set of computer programs to do real multilocus analysis. Lalouel and his colleagues knew what the problem was, knew how to solve it, and were doing it correctly. White's attempts at computer programming for multilocus analysis were, as he himself later freely admitted, pathetic. Fortunately (at least for White and his work at the University of Utah), White and Lalouel met and began to discuss the problem. Their meeting led to a professional and personal friendship that eventually brought Lalouel to Salt Lake City and White's lab at the University of Utah.

Why Utah? Why had Ray White gone there in the early 1980s to pursue his work on genetic mapping? Why did a prominent French researcher move to Salt Lake City from the City of Lights? Utah, in point of fact, is one of the best places in the world to carry out a project in genetic mapping. The reason has to do with the religious beliefs and practices of the Mormons.

The religion of the Church of Jesus Christ of the Latter Day Saints places a very high value on large, stable families, and on good health and dietary practices. The result is a strong and extended family structure for most of Utah's population. The Mormon religion also has a practice of what might be called "after-death baptism." One can have one's ancestors baptized as Mormons, made members of the Latter Day Saints church, and thus participate in all the spiritual privileges that go with being baptized Mormons—all after the fact, as it were. This practice has led to a huge Mormon interest in genealogy. The Latter Day Saint's genealogical library (whose records are supposedly kept in a thermonuclear-bombproof vault) is the best in the world.

And *this*, in turn, has naturally led to a keen interest in genetics. White found among Mormon families a great willingness to participate in the kind of study he wanted to carry out. And they enjoyed it. Salt Lake City, the state capital and the location of the University of Utah, is also the "Vatican" of Mormonism. That city, and all of the state, is a very special genetic resource.

Robert Sinsheimer pulled together the first formal meeting of researchers about the Genome Project in 1985.—Photo courtesy of the University of California, Santa Cruz.

Charles DeLisi almost single-handedly started the Genome Project ball rolling. He organized one of the first large conferences on the project, and continues to follow its development.—Photo courtesy of C. DeLisi.

There was one other reason for White's move to Utah. He was recruited there. The Howard Hughes Medical Institute, headquartered near Washington, D.C., asked White to run their center at the University of Utah. They promised him additional support for his gene mapping project, over and above the money he was to get from a modest grant from the National Institutes of Health. That would make it possible for White to scale up the project. White and his colleagues would then be able to identify enough genetic markers to begin mapping the human genome, and they would be able to do in in just five to ten years. It was, White would later say, an offer he could not refuse.

SANTA FE 1986

Charles DeLisi is one of the parents of the Human Genome Project. In 1985, this former physicist was bitten by the Genome Project bug. It was DeLisi's enthusiasm that led to the first truly large gathering of scientists and researchers to consider the Genome Project. Ironically, after all his letter-writing, phone calling, cajoling and arm-twisting, DeLisi himself did not attend the meeting.

Until 1988, DeLisi was the director of the Office of Health and Environmental Research (OHER) for the U.S. Department of Energy, the bureaucratic successor to the old Atomic Energy Commission (and later the Energy Research and Development Administration). From its very beginnings, the AEC had taken an active interest in human genetics. The reason was simple: The AEC was responsible for the building and testing of nuclear weapons and of nuclear power plants. Ionizing radiation from nuclear fission causes genetic mutations, and the AEC needed to know as much as possible about that process. Over the 30-plus years of its existence, the AEC and later the Department of Energy had developed an extensive expertise in human genetics. The people involved worked both at universities and at the agency's own national laboratories, including Los Alamos National Laboratory in New Mexico, Lawrence Livermore National Laboratory in California, and Oak Ridge National Laboratory in Tennessee. The OHER, DeLisi's office, oversees all the work by DOE-connected scientists that deals with research related to health and the environment, including genetics.

Before coming to OHER, DeLisi had been the chief of mathematical biology at the National Institutes of Health (NIH). He had given a great deal of thought to the mathematical and computational problems connected to the increasingly faster generation of data about DNA sequences. That led him to consider just what might be required to sequence an entire genome. DeLisi wasn't the first to think about such a problem. Nor was he the first to do anything concrete. It was Robert Sinsheimer (as DeLisi himself freely admits) who actually took the step of getting people together to brainstorm the concept of sequencing the genome.

In 1985, after coming to DOE, DeLisi began looking into the problem of understanding variations in people's innate tendency to contract certain diseases. (The process is technically called *diathesis*.) This had immediate bearing on the work of the DOE and the OHER. For example, some people will develop cancer when exposed to low levels of a toxic substance or radiation, but others will not. The person who pointed DeLisi in this direction was Mark Bitensky, then the leader of the Life Sciences Division at Los Alamos. Later that year, DeLisi was reading through a draft of a report entitled *Technologies for Detecting Heritable Mutations in Human Beings*, published by the Congressional Office of Technology Assessment (OTA). While reading the draft of the report, his concerns about diathesis variations and sequencing the genome came together.

As a physicist and former DOE director, DeLisi was used to thinking about large projects. He knew that a project to sequence the genome would be huge, and that it would inevitably have enormous spinoffs. One would be the development of technology for the quick comparison of large stretches of DNA from different people with and without specific diseases. He later recalled looking up from the report and thinking to himself: Wouldn't it be great if we could compare the genome of a child with that of its parents, base pair by base pair? What a wonderful tool that would be for studying genetic mutations. It was a concept of immense power, and it seized his soul and imagination. He started checking with people at Los Alamos and Lawrence Livermore to see if anyone was thinking along the same lines. One of the people he contacted was Mort Mendelsohn at the Lawrence Livermore National Lab (LLNL) east of San Francisco. He asked Mendelsohn what he thought of the idea of sequencing the human genome. He responded by telling DeLisi of Sinsheimer's workshop in Santa Cruz. DeLisi recalled Mendelsohn as being somewhat negative about a massive genome sequencing project.

DeLisi also knew about the work being done by Anthony Carrano, who worked under Mendelsohn at LLNL. Carrano was creating cosmid libraries—clones of DNA fragments. At that time Carrano had not thought about genome sequencing. However, laying cosmids out in a sequential order could produce a physical map of a DNA sequence. This kind of map would not depend on genetic markers as "signposts" along the way, as would Ray White's linkage maps being constructed at the University of Utah. Instead, such physical or cosmid maps would be just that: actual maps of the actual physical locations of the base pairs of a human DNA sequence. It would not be a complete sequence map of all three billion base pairs of the human genome; but it would be the next major step in that direction. White's maps were, in a sense, being constructed from the top down. Linkage maps would give the big picture of a chromosome, and then the details would be successively filled in to produce the total sequence map. Physical maps, on the other hand, would be constructed from the bottom up. They would begin with the details of a small part of a chromosome. The successive addition of more and more DNA clones would build the maps up into the ultimate map of the entire genome.

After talking with Mendelsohn at LLNL, DeLisi met the next day with other members of OHER, including the man who would later take his place, David Smith. DeLisi started making phone calls and writing letters. One of the people he contacted was Mark Bitensky at LANL. In December 1985, at DeLisi's urging, Bitensky began organizing an international gathering called "The Genome Sequencing Workshop." The meeting took place on March 3 and 4, 1986, in Santa Fe, New Mexico.

The 1986 Santa Fe Workshop was a both a beginning and an end. It was an end to more than thirty years of dreams and failed attempts to fuse mathematics and biology into a new scientific discipline that could map the genetic code. The development of new biological technologies in the 1960s and 1970s was what finally made it possible for researchers to see the creation of a genome map as a real possibility. Now, in the early spring of 1986, came a new beginning. The Human Genome Project was born at the 1986 Santa Fe workshop.

Attending the meeting were people who were already significantly involved in mapping parts of different genomes, including the human. A large contingent came from the Los Alamos National Laboratory. They included George Bell, Mark Bitensky, Christian Burks, and Walter Goad. Frank Ruddel came from Yale University; Charles Cantor, from Columbia University; David Smith, from the Department of Energy's OHER office in Washington, D.C.; Anthony Carrano from Lawrence Livermore National Laboratory; and Walter Gilbert from Harvard University. The Europeans were represented by Sydney Brenner, from the Medical Research Council in England, and Hans Lehrach of the European Molecular Biology Laboratory (EMBL) in West Germany. There were also researchers from Baylor University, Johns Hopkins University, the University of California at Berkeley, University of California at San Diego, University of California at San Francisco, the University of Colorado, the University of Florida, the University of Texas, the University of Virginia, and the University of Wisconsin. Other scientists came from the Brookhaven National Laboratory and the Cold Spring Harbor Laboratory on Long Island, the City of Hope Medical Center in California, the Fred Hutchinson Cancer Center in Seattle, and the Oak Ridge National Laboratory in Tennessee. Several corporations also sent people to the Santa Fe meeting, including Applied Biosystems in Foster City, California, Proteus Technology of Rockville, Maryland, and DuPont's New England Nuclear Research in Boston. In all, about 42 men and women attended the two-day conference.

The meeting had several objectives, DeLisi would later write. They included determining whether it was technically possible to sequence the human genome by the end of the century; the cost of such a project; its benefits to the United States and the Department of Energy, which would presumably spend a lot of money on it; and how such a Big Science project (big at least for the biological sciences) could be managed.

Much has since been written about the 1986 Santa Fe meeting, but none of it can quite capture the intensity and excitement of those two days in March. It surprised nearly everyone. DeLisi (who was not there) would later compare it to "those rare moments in the early phases of major ventures—such as the Manhattan Project at Los Alamos" (a poignant comparison, considering that the Manhattan Project's conclusion, the atomic bomb, gave birth to the agency that would sponsor the Santa Fe meeting) "or the conquest of space." If a single statement could capture the essence of "the spirit of Santa Fe," it would be the widely-quoted comment of the ebullient Walter Gilbert: "The total human sequence is the Holy Grail of human genetics."

Not surprisingly, to those who knew him well, Gilbert would soon present himself as the Parsifal of the quest for this particular grail.

THE DULBECCO ARTICLE

The next major turning point in the Genome Project took place just three days after the end of the Santa Fe conference. It was the publication of "A Turning Point in Cancer Research," an article written by Nobel Laureate Renato Dulbecco in the March 7, 1986, issue of *Science* magazine. Like DeLisi, Dulbecco had been at least partly inspired by a report on the Santa Cruz meeting.

The article's subtitle was "Sequencing the Human Genome." It began: "One of the goals of cancer research is to ascertain the mechanisms of cancer." Research into the causes and progression of cancer, Dulbecco wrote, had come full circle: from cell cultures to cancer-causing viruses, to the genes in such viruses which seem to cause cancers (called *oncogenes*, with the prefix "onco-" meaning cancerous), and back to the cancerous cells themselves and their genomes.

"We are at a turning point in the study of tumor virology and cancer in general," Dulbecco wrote. "If we wish to learn more about cancer, we must now concentrate on the cellular genome." The reason? It had become apparent to many cancer researchers in the last several years that "the state of the cellular genes is important for the effect of oncogenes." Viral oncogenes, in other words, may trigger the progression of cancerous tumors in some people but not in others. The reason appeared in part to be connected to the presence (or absence) of genes in the target cells. By the same token, chemical or viral carcinogens were known to induce cancers in some species of animals, but not in others.

The scientific studies which pointed in this direction dealt with the initial events in the development of cancer, Dulbecco noted, adding: "But natural cancers evolve slowly toward malignancy through many definable stages, in a process called 'progression,' which is the least understood but probably the most crucial phase in the generation of cancers and their many chromosomal abnormalities." A major gap in science' understanding was how an oncogene's activity is related to the events of progression. To understand that, however, would mean learning which genes in the cell were turned on, turned off, or mutated by the oncogenes.

As a result, wrote Dulbecco, "We are back to where cancer research started."

However, the situation in 1986 was quite different from that of the late 1950s and 1960s. Medical science had at hand new tools, such as genetic engineering, and more knowledge about viruses, oncogenes, and genetics. "We have two options," Dulbecco claimed: "either to try to discover the genes important to malignancy by a piecemeal approach, or to sequence the whole genome of a selected animal species."

The problem with the former approach was the extreme biological and genetic variety of different cancerous tumors, and the extreme paucity of cell cultures for the many different cell types present in a cancer. In a very real sense, each malignant tumor is unique because each human is genetically unique. The piecemeal approach of identifying the specific genes important in causing a tumor would be an almost infinitely vast undertaking, a new research project for each cancer victim.

The latter approach, sequencing the entire genetic code of a particular animal species, also seemed like a hopeless task. In fact, claimed Dulbecco, it would be much less massive than the piecemeal approach, would cost less in the long run, and would have some incredible medical benefits at least as important as the primary goal of curing cancer.

Wrote Dulbecco: "The [genetic] sequence will make it possible to prepare probes for all the genes and to classify them for their expression in various cell types at the level of individual cells. . . . The classification of the genes will facilitate the identification of those involved in progression."

As for which species should have its genome sequenced, the answer seemed obvious to Dulbecco: Us. Humans. Homo sapiens. If we want to understand human cancer, we should sequence the human genome. The genetic control of cancer seems to be different in different species, so it would be useless to sequence, say, the mouse genome in order to learn the genetic control of cancer in humans. One result of having the sequence of the human genome, Dulbecco added, would be that humans would become the preferred species of research on human cancer. Human cells would be grown in culture for this purpose, as they already were for other aspects of medical (including cancer) research. They could also be transferred into mice with suppressed or absent immune systems, and studied in a living creature other than a human.

A map of the human genome would also finally make it possible to study cancer on the molecular level. Knowing the existence and location of a gene would lead to understanding its molecular construction, and also knowing what protein it produces. That in turn would lead directly to the identification of the molecular agents that trigger cancer in humans. Knowing which genes in the human genome are involved in the progression process of cancer in humans would in turn lead to new ways of treating cancer. And that, concluded Dulbecco, "might lead to a general cancer cure if progression has common features in all cancers."

Dulbecco's focus in his March 1986 *Science* article was on the importance to cancer research of sequencing the human genome. However, he was not blind to some of the other consequences. A map of the human genome, and possession of genetic probes that could identify the presence of any particular human gene, he wrote, "would also be crucial for progress in human physiology and pathology outside cancer; for instance, for learning about the regulation of individual genes in various cells types. Many fields of research . . . would benefit. The identification and diagnosis of hereditary diseases or hereditary propensity to disease would be greatly facilitated. The knowledge would rapidly reflect on therapeutic applications in many fields." In other words, mapping the human genome would have a rapid and dramatic impact on treating and curing many illnesses, especially genetically caused ones.

Dulbecco's concluding paragraphs were prescient:

"An effort of this kind could not be undertaken by any single group: it would have to be a national effort. Its significance would be comparable to that of the effort that led to the conquest of space, and it should be carried out with the same spirit. Even more appealing would be to make it an international undertaking, because the sequence of the human DNA is the reality of our species, and everything that happens in the world depends on those sequences.

"Many practical and technical problems would have to be solved. A considerable improvement in the technology would be needed in order to shorten the time required. Increasing by 50-fold the present rate of sequencing would make it possible to complete the main task in perhaps 5 years with adequate manpower [sic]."

Dulbecco was right about the national and international emphases of the Genome Project. He was also correct about the need for technological improvements. But like most prophets, he missed the boat on other aspects. The Genome Project is not being pushed by the can-do spirit of the U.S. space program of the 1960s, in large part because of the budgetary and ethical concerns that surround this first "big science" project of modern biology.

And Dulbecco's primary goal, that of *sequencing* the human genome, is not the initial goal today. The 18 months of debate that followed the appearance of Dulbecco's article saw a significant shift in goals for the Genome Project. As one researcher later put it: "First map, then sequence."

CARRANO AND COSMID MAPS

One of the people that Charles DeLisi had met and talked with during his personal journey to the Genome Project idea was Anthony V. Carrano, the Genetics Section leader of the Biomedical Sciences Division at the Lawrence Livermore National Laboratory in California. He has long played a role in the development of techniques for separating, cloning, and categorizing DNA segments. In 1979, for example, Carrano and several colleagues described a method of measuring and purifying whole human chromosomes using a laboratory instrument called a flow-sorting cytometer. By the time he and DeLisi met in 1985, Carrano had been working for several years to develop a special kind of library—a collection of large DNA fragments, separated by the chromosome from which they come and arranged in overlapping sequential order. The DNA fragments were of two types, called *YACs* and *cosmids*. Cosmids, as noted earlier, are small clones of DNA from a genome. They run about 40,000 base pairs in length. YAC stands for Yeast Artificial Chromosome. YACs were developed by Maynard Olson at Washington University in St. Louis in the mid 1980s. They consist of very large pieces of DNA from another species spliced into the DNA of a form of yeast. YACs are large, from 200,000 to 600,000 base pairs long.

It is important at this point to make a distinction between genetic *mapping* and *ordering*. Mapping is the assignment of genes or *DNA segments* to specific chromosomes or chromosome regions. Ordering, on the other hand, is the establishment of the natural order of *large pieces* of a genome. This is done by arranging cloned DNA segments in the order in which they naturally exist in the genomic sequence. Carrano's work on a cosmid library had to do with ordering. He was using cloned fragments for several reasons. First, clones can be stored, reproduced, used in other genomic studies such as DNA sequencing, and shared with other researchers. Also, genes and other probes or markers can be located on cloned DNA segments, making it easy to isolate the genes themselves and determine the protein they make. Finally, when an ordered set of clones is actually linked to a genetic map, it can be used to establish which areas have the highest priority for sequencing. Researchers can select those clones which are most likely to contain a gene they are interested in.

Carrano was using two different approaches to create his cosmid library: the bottom-up approach and the top–down method. The bottom-up approach organizes sets of overlapping cosmids along the chromosome. These groups of cosmids Carrano calls *contigs*, from the word "contiguous" (adjacent or joining). The top-down approach uses the larger DNA segments from YACs to create a large-fragment map that closes the inevitable gaps that remain from the bottom-up method.

Both Carrano and DeLisi soon recognized the value of such a genetic cosmid library. Carrano and his colleagues began looking at ways of using it as part of the Genome Project effort. They decided to put together a cosmid map of human chromosome 19, as well as some other selected regions from the human genome. Chromosome 19 contains about 60 million base pairs, about two percent of the genome. At least 55 functioning genes or gene families have already been mapped to chromosome 19. They include a set of three genes that is involved in the repair of damaged DNA, and the gene responsible for a disease called myotonic dystrophy. Carrano constructed two cosmid libraries for that chromosome, using DNA from hamster-human hybrid cells, and chromosome 19 separated using a flow-sorting machine. A third cosmid library consisted of 600,000 base pairs of DNA from part of chromosome 14, inserted into a yeast artificial chromosome supplied by Maynard Olson at Washington University.

They worked on a "proof-of-concept" experiment to put together a series of contigs that would cover a small part of chromosome 19. By the end of 1988 Carrano and his colleagues had succeeded in creating an ordered set of cosmids that traversed about 140,000 base pairs of chromosome 19, about two-tenths of a percent of the chromosome.

They also carried out an experiment which ordered the set of chromosome 14 cosmids contained in the YAC. They were able to assemble the 82 cosmids they had into three contigs and six "islands" (isolated cosmids). It took the team less than two man-weeks to finish the work on the 600,000 base pair YAC.

COLD SPRING HARBOR 1986

When Paul Berg and his colleagues had worked with the lambda virus in the 1970s, they knew only that its DNA was about 50,000 base pairs long. About half of lambda's DNA was involved in expressing so-called "early" genes and the other half with the late genes. They were able to determine which genes were which by using the Eco RI restriction enzyme. Eco RI acted as a marker or flag pole. Go *this* way from the Eco RI cutting site, and these are early genes. *That* way lie the late genes. Later, a researcher named Daniel Nathan discovered other restriction enzymes which could cut the lambda virus's DNA into eight to ten large fragments. Scientists were then able to make the first physical map of the lambda phage. Several of the fragments contained the early gene regions, while others held the late regions. This was a tremendous breakthrough. Researchers could now isolate the different DNA fragments of the lambda phage and use them as genetic probes. The real power finally came with the development of a sequence map for lambda phage, when the entire 50,000 base pairs were sequenced. The level of sophistication of manipulating the DNA increased by many orders of magnitude. Researchers could now read off where the genes were; find the exact DNA sequences preceding or following a gene; know exactly where to cut the DNA with restriction enzymes; and understand how to splice the viral genome back together. They had gone from having some vague information about the lambda virus's genome, to a physical map, to the total sequence. Later, the same process took place with other viruses as their sequences were mapped.

However, the first that Berg heard of the idea of seriously sequencing the *human* genome was when he read Renato Dulbecco's article in *Science*. Berg thought that Dulbecco's arguments about the benefits of sequencing the human genome were vague and unconvincing. Nevertheless, he was interested. He later recalled that his response to the Dulbecco article was essentially: "Well, there's no question that we need to understand the human genome. Genetic functions, gene arrangement, subtle genetic sequences, the coding for regulatory signals, and other things we can't imagine—they'll finally become apparent when we have the sequence."

James Watson had organized a scientific meeting on the genome at the Cold Spring Harbor Laboratory on Long Island. The title of the conferences was "The Molecular Biology of *Homo sapiens*." It was set for May 28 through June 4, 1986, about two months after the Dulbecco article in *Science* and the Santa Fe meeting, and a year after Robert Sinsheimer's Santa Cruz gathering. Berg wrote to Watson and told him that it seemed reasonable that a meeting on the human genome should have at least a little time set aside to discuss the Dulbecco proposal.

At this point Berg was totally unaware that there had already been several meetings about the genome proposal, including Sinsheimer's get-together at University of California at Santa Cruz, and the Santa Fe meeting engineered by DOE's Charles DeLisi. Indeed, he had no idea that the Department of Energy was already moving aggressively to capture the "high ground"—and the lion's share of government funding— for a genome initiative. When he finally arrived at Cold Spring Harbor, Berg found that Watson was way ahead of him. A full afternoon session was scheduled to discuss the Dulbecco genome proposal. Watson asked Berg if he would mind co-chairing the session, along with Walter Gilbert. Berg agreed. He thought that, in his words, "a small group of genome nerds would show up to talk about it, and everyone else would go to the beach."

Berg's small gathering of genome nerds turned out to be the most heavily attended session of the entire week-long meeting. The attendees filled the entire auditorium at Cold Spring Harbor. Berg had really not been prepared to chair the session, so he went off by himself to think about it for a bit. He knew that many people would be concerned about the financial aspect of a human genome project. Would such an undertaking cut into federal funding for other people's projects? Researchers get testy and defensive when they perceive their financial lifeline being threatened. Berg felt that the best strategy to follow for the afternoon session would be to completely dismiss the funding aspect of the project. He would propose that the gathered multitude instead play a game of "what if"? Imagine (he would say) that someone gives us all the money we need to carry out a human genome project of the kind that Dulbecco has suggested. Just assume we've got all the money we need to do it. Now what? Is it worth knowing the sequence in the first place? How would we do it in a sensible way? What is an appropriate strategy? What would we learn from sequencing the human genome?

Berg never got the meeting onto that track. Almost from the beginning of the session, people were jumping up and voicing their violent opposition to any such undertaking or project. What terrible things it would do to science! Lousy quality of scientific work! Big Science versus little science! *I'll lose my funding*!

Though he and Walter Gilbert are long-time colleagues and fellow Nobel Laureates, Berg felt that it didn't help anything that Gilbert was co-chairing the session. Then Gilbert started talking about carrying out a genome project as a "blind sequencing" project. Essentially, that meant cutting pieces of human DNA at random, sequencing them, and then arranging them in sequential order. Such a project would cost about one dollar per base pair. The high point (or low point, in Berg's estimation) was when Gilbert went up to a blackboard and wrote the figure "$3,000,000,000". That scared people.

The atmosphere at the Cold Spring Harbor symposium was far chillier than that at Santa Fe, and it had nothing to do with altitude. The participants at the Santa Fe meeting had pretty much focused on the science of a genome project. Those at the afternoon session co-chaired by Berg got caught up in the politics and economics of it. Gilbert's three billion dollar figure sent shivers down the spines of many participants. MIT's David Botstein (who would later become a vice-president of the bioengineering giant Genentech Corporation) reminded those present that they were babes in the woods when it came to playing hardball scientific politics. If they tried to get the feds to kick in massive amounts of money for a genome sequencing project, he warned, they would end up only getting their own research funds cut.

Maxine Singer, then at the National Cancer Institute in Bethesda, Maryland and later the director of the Carnegie Institute, offered another point of objection. She did not see any great urgency in the matter. There were better ways of learn-

Maxine Singer, President of the Carnegie Institute of Washington, has been involved in the deliberations about the Genome Project from its beginning.—Photo courtesy of the Carnegie Institute of Washington.

ing about the human genome than jumping into a full-fledged genome sequencing project. She suggested an approach that combined biochemistry and genetics with mapping procedures.

David Baltimore of the Whitehead Institute in Cambridge, Massachusetts, one of several Nobel Laureates in attendance, greatly opposed the very idea of launching a massive project to sequence the human genome. Better, he said, to come up with a physical map showing the relative locations of the genes. A sequence map could later be generated for the piece of DNA that a researcher might be interested in. He later declared himself to be "shivering at the thought" that a drive to create a sequencing project might be gathering momentum.

Sydney Brenner of the Medical Research Council's Laboratory of Molecular Biology in Cambridge, England, was (and still is) one of the major British figures in the Genome Project.

He agreed with Baltimore that a full-fledged effort to sequence the genome was premature at best. He, too, supported the idea of beginning with constructing a physical map rather than trying to blindly sequence the genome. And he planned on making a go at it himself by the end of 1986.

Not everyone at Berg's afternoon gathering was opposed to a genome sequencing effort, however. Walter Bodmer of the Imperial Cancer Research Fund in London said he thought it was the most exciting human endeavor in scientific history. And David Smith of the Department of Energy's Office of Health and Environmental Research reminded the conferees of the great enthusiasm seen at the Santa Fe meeting a few weeks earlier. That enthusiasm, he said, had made a big impression on the DOE higher-ups. The department was setting up advisory committees, he added, and was going to ask for funding of a set of genome project research proposals. Although nothing was funded yet, the money might be in the order of $20 million over three years. DOE, Smith said, would concentrate on improving the technology of genetic sequencing, including automatic sequencing machines; developing a physical map of the genome; and devising new and more powerful methods of data-handling.

David Smith's comments touched on a raw nerve at the Cold Spring Harbor meeting. Many in attendance were very unhappy with the deep involvement of the Department of Energy in a potential genome project. Despite its decades of experience in genetic science, its ability to manage massive scientific and technical projects, and its constellation of world-class national laboratories, they felt intuitively that a biological science project should be led by some agency with biological roots—say, the National Institutes of Health. The Energy Department's deep involvement in weapons research may also have troubled some participants. A more purely scientific organization, such as the National Science Foundation, would be much more palatable as the center for the Genome Project.

The response from Smith was essentially: "A genome project would be more of an organizational challenge than a scientific or technical one. The DOE knows how to do Big Science, and the NSF and NIH do not." It was an argument that did not go down well with most participants.

In the space of three months, one season, the Genome Project seemed to have gone from spring right into the dead of winter.

5

Small Science, Big Science

Darest thou now, O soul,
Walk out with me toward the unknown region,
Where neither ground is for the feet nor any path to
follow?

WALT WHITMAN
Leaves of Grass

O NE OF THE BIGGEST controversies within the scientific community over the Genome Project has been the issue of Big Science versus Small Science. It came quickly to the fore in 1986, just as the momentum for the Project began to build. Indeed, the controversy threatened to end the Genome Project before it had even begun. The issue was more or less resolved by the end of 1986, but it continues to simmer.

What exactly is meant by "Small Science" and "Big Science"? In particular, how do these phrases apply to the sciences of biology and genetics? And what influence will the Genome Project have on them?

IMAGE VERSUS REALITY

Most of us carry imaginative ideas of the scientific process that we derive mainly from movies and science-fiction novels. They come in two versions. The first is of the lone scientist working in a basement laboratory, cobbling together (1) a spaceship that takes him to Mars, or (2) a hideous monster/virus/clone of Adolf Hitler/robot which terrorizes the world. The important point here is the picture of the lone scientist. This is the popular image of Small Science. The second image is that of the team player scientist who works in a huge, antiseptic laboratory run by a huge corporation/the government/a giant university. He (scientists are always "he" in these images) helps in the creation of (1) and/or (2) mentioned above. This is the popular image of Big Science.

Like most of our mental pictures taken from movies or prose fiction, these two images of science are distorted versions of reality. Of course, not all written or filmed depictions of science are so severely distorted. For example, *The Double Helix*, by James Watson, is the account of how he and Francis Crick discovered DNA's structure. *Timescape*, by physicist Gregory

Benford, a fictional story of the discovery of messages from the future, presents an accurate picture of how physics is actually done. The truth is often duller than fiction, and that is frequently the case with real science.

For the truth is this: The basic unit of contemporary scientific research, whether in the physical or biological sciences, continues to be the small group or team. At the center of the team is a professor or senior researcher. He (and it still is usually a "he") usually does not do much actual work at the laboratory bench, at the telescope, on the particle accelerator's magnets. Rather, he is the overall supervisor or mentor. The rest of the team is made up of a few graduate students, one or two post-doctoral fellows, and sometimes a few technicians. The entire team will probably consist of no more than a dozen people—twenty at most. The environment in which the research group works depends a great deal on the professor. It can be conservative, highly controlled, and tightly structured. Even better, however, the environment will be loosely structured, fairly casual, with lots of room for individual initiative on the part of the students, and plenty of intellectual and emotional stimulation. For the real point of science is not so much to find new answers. It is to discover new questions, and in the process uncover interesting answers. If the answers are useful in a technological sense, so much the better. But "usefulness" is not the prime goal of the scientific process. This process is best carried out in an environment that encourages question asking, even if the questions seem positively outrageous. Keeping the size of the group fairly small allows considerable intellectual interaction.

The result is that nearly all the great science that has been done in the last century, not to mention the last 25 years, has been Small Science. It has given us the theories of relativity and quantum physics, the ideas that led to the creation of the nuclear reactor and the thermonuclear bomb. It produced the first image of what the DNA molecule looks like, and gave birth to molecular biology. It created the laser and thus the medical, metallurgical, and military industries that depend on that coherent beam of light. Small Science has given us the semiconductor and the huge industries that have grown up around it. Jobs and Wozniak would not have been able to cobble together the first Apple I computer in a suburban California garage were it not for the existence of the computer chip. Very Small Science—two researchers working in an IBM laboratory in Switzerland—discovered high-temperature superconductors, and won the Nobel Prize one year later.

And it was a small team of researchers who came up with the idea for a machine which would change the way biologists sequenced DNA.

THE AUTOMATED DNA SEQUENCER

In 1985 Lloyd Smith and Leroy Hood of Caltech began seriously considering ways in which the tedious manual method of sequencing segments of DNA could somehow be automated. With the assistance of several other researchers and graduate students, including the brothers Michael and Tim Hunkapiller, Smith developed an ingenious method which used fluorescent dyes, a laser, and a computer to speed up the standard Sanger method of sequencing DNA fragments. The team also had considerable support from a company located near San Francisco, Applied Biosystems, Inc. (ABI). Caltech and ABI then struck an agreement which gave the company the exclusive license to make commercially available the DNA sequencers (and several similar machines which sequenced proteins and automatically constructed DNA and protein molecules).

The method for automated DNA sequencing developed by Smith and Hood and their colleagues is based on the detection of fluorescence from DNA fragments which have been labelled with special dyes. In one sense this is similar to the manual method of sequencing DNA, which often uses radioactive molecules attached to the DNA fragments. The difference, of course, is that fluorescent dyes are not radioactive. This meth-

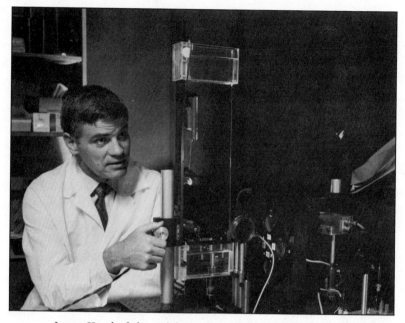

Leroy Hood of the California Institute of Technology. He and his colleagues developed the automated DNA sequencer later commercialized by Applied Biosystems, Inc. Hood is deeply involved in the Genome Project as a researcher, advisory board member, and constructive critic.—Photo courtesy of California Institute of Technology.

ods thus avoids all the potential health and environmental problems associated with using radioactive markers on the fragments. The new method uses four different dyes. Each dye molecule attaches itself to one of the four different DNA base pairs—adenine (A), cysteine (C), guanine (G) and thymine (T). The dye-tagged DNA fragments are allowed to percolate through a glass tube filled with a special transparent gel. The smaller fragments travel more quickly than the larger ones. More importantly, each dye fluoresces at a different color. The dyes are excited to glow by the beam of an argon-ion laser. The laser is beamed through the mixture as it nears the bottom of the gel-filled tube. The light emitted by each dye-marked fragment is focused by a special lens through a filter and onto

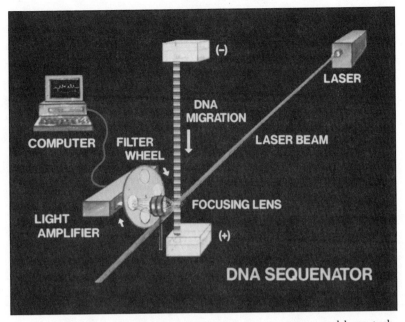

The DNA sequencer is a machine using computer and laser technology to sequence pieces of DNA automatically.—Courtesy of California Institute of Technology.

a photomultiplier tube, which increases the strength of the light. The tube then turns the light falling onto it into digital signals which are directly detected by a computer. The computer in turn is running a program that is able to identify the light as coming from an A, T, C, or G base pair. The software also automatically arranges the fragments in their proper order, and thus sequences the DNA fragment.

The automated DNA sequencer, now commercially available from ABI, is speeding up the rate at which researchers can sequence segments of DNA. Work that used to take a week or more can now be done overnight. The automated sequencers are also reducing the costs of sequencing DNA. When Walter Gilbert estimated that sequencing the genome would cost $3 billion, it was with the assumption that sequencing would cost $1 per base pair. The machines developed by Hood and Smith have cut that cost to less than fifty cents per base

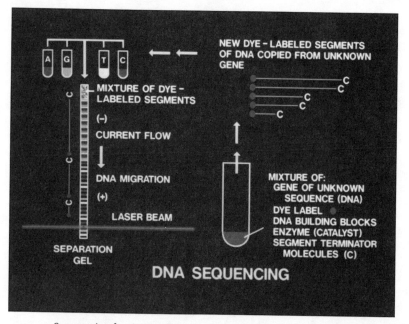

DNA SEQUENCING

Sequencing by ABI's DNA sequencer begins with a mixture of DNA with an unknown sequence of bases. Labels consisting of molecules of four different fluorescent dyes are attached to the ends of different DNA segments. Segments which end with an A (adenine) have one color attached, those ending with T (thymine) another, and so on for each of the four bases of DNA. A mixture of the dye-labeled segments is then put into a tube containing a special gel, and an electric current is applied to it. The DNA fragments migrate through the separation gel, with different pieces moving at different speeds because of the current. At the bottom, a laser beam excites the label and causes it to fluoresce. An electric eye detects the color and passes the information to the computer, which arranges the As, Ts, Cs, and Gs in their correct order.—Courtesy of California Institute of Technology.

pair. Further developments and still newer generations of automated sequencers will soon drop the cost of sequencing DNA to ten cents per base pair—or less.

CONSENSUS AND OFFENSES—JUNE 1986

If there was any consensus reached at the 1986 Cold Spring Harbor meeting, it was a consensus of reservations about leaping into a massive genome sequencing project. The emphasis of the reservations was on *sequencing*. Two kinds of objections surfaced, as Eric Lander of the Whitehead Institute later told *Science* magazine. The first had to do with the overall strategy leading to a sequence map of the human genome. Lander noted that the human genome *would* likely be sequenced by the year 2000. What was important was the route taken to get there. Researchers now had the ability to begin sequencing the genome, and eventually produce a complete sequence map with the location of every base pair. Such a project, though, would be on the scale of NASA's space station—at least relative to biology. That science, Lander pointed out, was very good at devising new laboratory methods and machinery, and in carrying out small-scale projects. Big Science was something still unknown in the field. A gargantuan genome sequencing project of the kind envisioned by people like Charles DeLisi and Walter Gilbert, said Lander, would literally change the nature of biological research. And Big Science structures for biology, as immovable as all such bureaucracies are, would only damage the science in the long run. Indeed, most of the great discoveries and developments of new techniques in biology have been serendipitous, and not the result of some planned major project.

If the first objection had to do with strategy, the second dealt with tactics. "In one sense, the sequence is trivial," Lander told *Science*. Other researchers agreed with that assessment. A particular kind of physical map of the genome, on the other hand, would not be trivial at all. It would have immediate practical benefits for medical science.

One way of envisioning this is to imagine the usefulness to a traveller of three kinds of maps. The traveller is on vacation, and she is driving from New York to Seattle for the first time. The first map is a map of the world. The second map is a street map of downtown Seattle. The third map is a AAA highway map of the United States. The world map is like current cytogenetic and restriction maps of the human genome. It does not give the traveller enough information to help her plan her trip. It may not even show the major roads connecting New York to Seattle. In the same way, cytogenetic and restriction maps represent a very low-resolution image of the genome, and give the location of different genes on the chromosomes. Researchers cannot use cytogenetic maps to find and remove manageable sections of interesting DNA.

The second map, that of downtown Seattle, gives a great deal of information about highways, streets, and even alleys. The traveller may well want such a map, for it will come in handy—after she gets to Seattle. Its great detail, however, is useless at the moment. In the same way, a sequence map with its extremely high resolution of the genome will have its uses. But they are not yet apparent to many researchers, nor is the information to be found in such a map even usable at this point.

The third map, showing major highway routes across the United States, is the one that the traveller will find most useful. Similarly, a cosmid map of the genome will give researchers enough resolution to find DNA sequences with both medical and purely scientific interest. Such a map of the genome could be constructed from overlapping cosmids about 40 thousand base pairs long. Lander, Brenner, Baltimore, Singer, and others at the 1986 Cold Spring Harbor meeting much preferred a focus on the construction of such a cosmid map of the genome.

To Paul Berg, too, it didn't make sense at that time to talk about sequencing the genome. It would be much better first to begin upgrading current biological and computer technology, and invest in new technology. However, the people at the session got so upset over Gilbert's blind sequencing idea and the figure chalked on the blackboard that they rebelled. Berg later remarked that he was appalled and outraged that no one wanted to talk about science at a scientific meeting. "Everyone wanted only to talk about petty politics, and 'protect my ass,' and 'what I am doing is important and this is not,' " he recalled bitterly. He was also offended by the arrogance of many of his colleagues who said basically, "Hey, we don't need to know any of this. Most of the human genome is in the introns, anyway, and it means nothing. Forget it."

Introns are large segments of DNA found in mammalian and other eukaryotic cells. Unlike *exons*, those segments of DNA that contain the coding for genes, introns appear to code for nothing. However, the operative word here is *appear*. No one really knows for sure that introns are really silent, or contain nothing but "junk DNA," or are merely "spacing" in the DNA of higher order animals. Berg was doubly angry about this argument against sequencing the human genome. These scientists were simply assuming that much of the human genome was "junk" and that sequencing it would be a waste of time. More than that, there was the hypocrisy involved. Many of the people who dismissed the idea of a genome project because "we don't need to know any of this" were frequently the same people who would interrupt conferences to talk about "how little we understand about the process of the development of a trillion cell organism from a single cell." How could anyone conclude that there is data in the genetic sequence that "we don't need to know"?

Berg was also offended by the notion that we need to know only the structure of the genes and what regulates their expression, the way they produce proteins. The genome does much more than that. The DNA in the chromosomes is packaged in supercoils into an incredibly tiny volume. Somewhere in the genome are the instructions to carry out that compactification in the nuclei of cells. What's more, the genome must also be replicated, segregated, and recombined. It is involved in many different kinds of what Berg calls "genetic chemistry." We know nothing about how the genetic sequence directs those processes. For example, it is now known that the way a DNA molecule *physically bends* is important to its expression as proteins and peptides. The DNA sequence itself must contain all the data to direct this. Where it is, and how it works, no one yet knows.

"We are in for many surprises," Berg later commented. "I will bet anybody anything—and I have—that we will see a succession of surprises."

He is almost certainly correct. The whole history of our unfolding understanding of the genetic code is rife with surprises. Introns are one good example. It was a totally unexpected discovery that bacterial genes would be completely different from mammalian genes. For bacterial genes do not contain introns. Mammalian genes do. Paul Berg himself was one of the discoverers of introns. He freely admits that it took a long time to recognize that they existed—because he and others didn't believe they *could* exist. It is a clear example of one of the devastating mental traps that science falls into again and again. An assumption about the nature of reality becomes so embedded in the scientific worldview that some facts and truths become invisible.

Another example of the surprises that have unfolded in genetic science also comes from Berg's personal experience. The SV40 virus is capable of processing a transcript of part of its DNA in two different ways, so as to make two kinds of genes from the same DNA sequence. Not long after this discovery, Berg gave a talk at the Salk Institute in San Diego, California, in which he predicted that this form of "alternate splicing" would prove to play a significant role in the regulation of gene expression in mammals. Francis Crick—who by this time had left England to take a position at the Salk Institute—disagreed with Berg's prediction. Crick gave the audience several cogent arguments about why Berg's prediction was bunk. For example, Crick said, viruses have only a small amount of DNA, so they need to be flexible in the regulation of gene expression. But mammals have lots of DNA in their nuclei, so they have no need to be so flexible. Berg bet Crick two cases of wine that Berg was right. Apparently many in the room also made wine bets with Berg. He took them. And then they all waited.

It took about eight years, but Berg collected on the bets.

Berg has also bet another scientist, Robert Weinberg, a case of wine that all the "nonsense" in the introns—the so-called "junk DNA"—is actually very important, and will code for significant information for the organism. Berg is convinced he will collect on this bet, too, adding to what must already be an extensive wine cellar.

THE BIG SHIFT

The initial driving force of the Genome Project had been the concept of getting a complete readout of the human genome—the spelling out, in order, base pair by base pair, of all three billion base pairs in the human sequence. The Santa Cruz meeting in 1985 and the Santa Fe meeting in March 1986 had convinced the scientists in attendance that such a Big Science project was feasible. The Cold Spring Harbor meeting in May threw cold water on the enthusiasm for such an approach. But another crucial turning point for the Genome Project took place only a few weeks after that. It was a meeting on July 23, 1986, at the National Institutes of Health in Bethesda, Maryland.

Entitled "Informational Forum on the Human Genome," the meeting included Japanese and European representation as well as the now-familiar American and British faces in the Genome Project debate. That signaled the growing international interest in the idea. More immediately, though, it was at this gathering that the idea of mapping the genome decisively replaced sequencing as the primary goal of the Project. The actual turning point probably took place in the middle of a talk by David Smith of the DOE's Office of Health and Environmental Research (OHER).

Every scientific meeting and retreat up to this point had not neglected the idea of mapping the genome as opposed to sequencing it. Nearly everyone recognized that the different types of genetic maps—cytogenetic, restriction, and physical—would give researchers and medical doctors invaluable information about the genetic code. Nearly everyone acknowledged the importance of continuing to create restriction maps of the kind being worked on by Ray White and his colleagues, and of devising physical or cosmid maps of different chromosomes, as Tony Carrano at Lawrence Livermore was working on. But it was the bold concept of totally *sequencing* the genome that had generated all the excitement.

One reason for the flowing adrenaline lay in the truth of Walter Gilbert's comment at the 1986 Santa Fe meeting. For many biologists and geneticists, the total sequence of the human genome really *was* "the Holy Grail" of genetics. The realization that it was actually possible to obtain the sequence was, for many, akin to a religious experience.

A second reason lay in the scale of the project. To sequence the genome would probably take $3 billion and some 30,000 person-years of effort. For the first time, biology would have a Big Science project of its own. Biologists would be in the same league as physicists and astronomers, at least in terms of federal budgets and the accompanying political clout. That was a very attractive vision to some.

To others it was a nightmare, and they had spoken loud and clear at Cold Spring Harbor. David Botstein said that blindly sequencing the genome would only indenture biologists to a mindless, boring task. James Watson told the NIH gathering that he had qualified support for the idea, although nearly everyone else at the Cold Spring Harbor meeting was opposed to it. The reason, Watson was quoted as saying, was that many of the researchers in attendance were young. They saw the prospect of a gargantuan Big Science genome project, consuming billions of dollars and years of effort, as a threat to their own research proposals. Paul Berg's memory of events, though basically the same, differed in one respect. His impression had been that even Watson did not support the Big Science sequencing idea. It appeared that only Gilbert, David Smith, and perhaps a few others at Cold Spring Harbor were on the Big Science bandwagon.

That bandwagon was bushwhacked at Bethesda in July 1986. According to an account in *Science* magazine, it happened as David Smith of OHER was speaking. Smith was attempting to present some slightly more modest economic figures for a sequencing project. He told the gathering that advances in automated instruments for sequencing DNA would cut the size of the effort by a hundredfold, to 300 person-years, and the cost by tenfold, to a mere $300 million. At that, Walter Bodmer (who had expressed some enthusiasm at the Cold Spring Harbor meeting) interrupted Smith and said, "But this doesn't address the issue of mapping which we know to be so important."

At that point in the meeting, and in the history of the Genome Project, it was clear that priorities had shifted. Months earlier, when he had first heard of the Genome Project idea, Paul Berg had felt that it would have to proceed in such a way that benefits would emerge constantly, at every stage. The taxpaying public would not sit and wait patiently for Walter Gilbert's three billion base pair, $3 billion "grail," to be handed to them. For that reason, Berg thought it would be necessary to stage the project in such a way that practical benefits would quickly emerge. His advice, offered to all who would listen, was: Don't worry about getting the entire sequence right now. Fund some mapping projects first, so that some practical benefits—the location of the cystic fibrosis gene, for example—would emerge. The development of physical or restriction maps would thus be a logical first goal. Just as important would be the process of putting in place the technology for computerized databases to handle the information from such restriction libraries, and to store the DNA fragments and probes developed in the mapping process.

And now Berg's idea of a genome project structured to produce quick and usable results was coming to the fore. *First* map, *then* sequence. Sequencing the genome was important, yes. But a map of the genome, and certainly detailed maps of different chromosomes, were more important. His friend

James Watson was later quoted in *Science* as saying that "getting a library of overlapping cosmids, a map, is acceptable. There should be more urgency for [this] than there is, because of the great benefits for genetic diseases and common diseases. We can't think intelligently about [the total sequence] until we have [a physical map]."

With the conclusion of the NIH meeting, the Genome Project could be seen as having at least two phases. Phase One would be the effort to create a physical map of the genome, using overlapping cosmids about 40,000 base pairs in length. This first phase would take about three years and about $20 million.

Phase Two would encompass the sequencing of the genome to produce what Victor McKusick of Johns Hopkins University has called "the ultimate map." This sequence map could be the "spelling out" of the entire human genome, with the actual locations of all three billion base pairs. On the other hand, the ultimate map might contain the base pair sequences of only those parts of the genome deemed "interesting" or "important." No one would hazard a prediction as to how long Phase Two would take. The guesses ranged from five to twenty-five years, or more, depending on how much money and manpower was available, and how much of the genome would actually be sequenced.

Walter Gilbert has suggested there is a Phase Three to the Project, which he has called "the complete understanding of all human genes." If the duration of Phase Two was indeterminate, that of Phase Three was positively indecipherable. The human genome may contain as many as 100,000 genes. To find out what proteins they make, what those proteins do, how different genes and gene clusters interact to create an adult human being from a single fertilized egg—who knows how long that would take?

BIG SCIENCE—THE POSSIBILITIES

The Genome Project has not been without its detractors and critics within the biological community. Early in its still-short

Nobel Prize winner David
Baltimore, at first an early
skeptic about the Genome
Project, is now a moderate
supporter of the effort.
—Photo by M. Lampert/
Boston.

history, one of the most vocal critics was Nobel Laureate David
Baltimore, director of the Whitehead Institute of Biomedical
Research. (Interestingly, the Whitehead Institute is also home
to mathematician Eric Lander, one of the biggest boosters of
the Genome Project.)

One of Baltimore's concerns has been about money, or the
lack of it. In this way he has echoed the concerns voiced so
vigorously at the 1986 Cold Spring Harbor meeting. The federal budgetary pie for medical research has never been a large
one, and it has not gotten all that much larger in the era of
AIDS. Indeed, many both within and outside the medical community have faulted the federal government for not giving
AIDS research enough funds. Given that state of affairs, it is
not surprising that not all other scientifically important
biomedical research gets adequate funding, either. Baltimore
has said that scientists are naive if they think the Genome
Project will not compete with other biomedical projects for

funding. His particular concern has been that AIDS research will come the short in any Congressional funding battle.

Baltimore may have touched on a salient point in one respect: concern about Congressional interest in the Genome Project. In recent years the politicians on Capitol Hill have shown a disconcerting interest in "micro-managing" more and more agencies, departments, and programs. This propensity, combined with the traditional practice of pork-barrel politics, could cause serious problems for the course of the Genome Project. One can imagine Congress, for example, appropriating large sums of money for specific genetic research projects, or for labs that happen to be located in some congressional leader's home state. If that should happen, Baltimore has warned, then the decision about which Genome Project labs or programs get funded may come to depend on raw political power, rather than on scientific merit. Indeed, many scientific researchers and political analysts feel this is already taking place in the Big Science fields of physics and defense research. Could Big Science biology be the next field to suffer from such practices?

Several years after the 1986 Cold Spring Harbor conference, David Baltimore was adamantly insisting that he was never really opposed to the Genome Project. This was inspite of direct quotes to the contrary in the news report section of *Science* magazine. But even if Baltimore *has* changed his mind and now supports the Genome Project as Big Science, his reported criticisms and concerns were representative of many others'.

On the other hand, Leroy Hood of the California Institute of Technology feels that the Genome Project as a Big Science effort is not only possible, but desirable. He sees the Genome Project as biology's equivalent of the Apollo moon landing program of the 1960s. And that is the way it should be "sold" to Congress and the American people. What's more, unlike the Apollo project, the Genome Project can and will have quick and immediate benefits for people. Discovering the location of genes for different genetic illnesses will lead to the development of diagnostic tests, and eventually to vaccines and cures. That's something people understand and identify with, since it touches them personally. It is certainly something that *politicians* can identify with. One prominent scientist who supports the Genome Project wryly noted that mapping the location of the genes for Alzheimer's disease, and sequencing them, would be a major political boost for the Genome Project.

Lee Hood has also pointed out another positive aspect of the Genome Project as Big Science. Even small laboratories would not have to spend a lot of time and money isolating and sequencing pieces of DNA. The information uncovered by the Genome Project will eventually be available online, in several large computer databases. That would make it possible for small labs, and even individual researchers, to download the data containing a DNA sequence, a map of some particular gene, or part of a chromosome. That will make it possible for virtually any scientist to make a contribution to genetics. Such opportunities will not be limited to those who work for giant drug companies, universities, or the federal government. The Big Science of the Genome Project, in other words, could make biology and genetics even more like "cottage industries" than they are now.

Concerns about congressional micro-management and political meddling assume that biologists are unwilling or unable to learn how to play the federal funding game. Clearly, physicists and astronomers have learned how to maneuver their Big Science projects through Congress. Space telescopes get built and launched into orbit. Superconducting supercolliders get funded. When University of Utah chemists Martin Fleischmann and Stanley Pons claimed, in 1989, that they had discovered a cheap way to create controlled nuclear fusion, one of the first consequences was a visit to their labs by prominent members of Congress and the Bush administration. Their claims have not proven out, but their political footwork was illuminating.

In fact, the decision by James Wyngaarten of the NIH to bring James Watson in as head of their Genome Project initiative indicated that the biology community had enough political smarts to make a go of a Big Science Project. Watson is not only a Nobel Laureate and a scientist of considerable creativity, but he is also known as a consummate manager and player of the political game in science. The Department of Energy, the other major government agency involved in the Genome Project, has long had expertise in managing large laboratories and Big Science projects.

When push comes to shove, the geneticists and biomedical researchers of the Genome Project do have two important things going for them: the federal government *is* interested in the Project; and the cost of sequencing the genome is going to drop. In 1986 Walter Gilbert was writing "$3 billion" on a chalkboard at Cold Spring Harbor. He was right. A year later Leroy Hood was estimating that it still cost one dollar per base pair to sequence large stretches of DNA by hand. But even then, Hood and his colleagues at the California Institute of Technology had invented automatic DNA sequencers, machines which were cutting the cost of sequencing by a third. It was clear even then, as the Genome Project was beginning to move forward as a distinct technological entity, that the cost of sequencing the genome was going to come down. Most analysts now believe that it will cost about ten cents per base pair by the early to mid 1990s. Continued developments in computer technology, including parallel processing computers, neural networks, and new generations of high-density-memory chips, will eventually drop the cost to pennies per base pair.

In the debate over Big Science versus Small Science, it may also be worthwhile to remember the title of a Buddy Miles jazz-rock song from two decades ago: "Compared to What?" The Genome Project could involve spending $3 billion over fifteen years to map and sequence the human genome (along with several others). That's an average of $200 million per year. Is that really Big Science? Compared to what? Watson and Crick's discovery of the double helix nature of DNA (in 1953) may have cost a few tens of thousands of dollars. That would include their salaries or stipends, and the cost of the cardboard Watson used to make their first model of the molecule. The Genome Project is Very Big Science compared to that. It cost the National Aeronautics and Space Administration (NASA) about a billion dollars over seventeen years to send the Voyager 2 space probe on a "Grand Tour" of the outer solar system planets including Jupiter, Saturn, Uranus, and Neptune. Even compared to that, the Genome Project is Big Science.

On the other hand, the U.S. Space Station program could cost more than $40 billion over a ten-year period. NASA spent $24 billion in about ten years to send two dozen men to the moon and put twelve of them on the lunar surface. The Superconducting Supercollider, a gargantuan particle accelerator to be built in (actually, under) Texas, will probably cost about $6 billion, twice as much as the Genome Project, and produce "nothing more" than previously unseen subatomic particles with names like the top quark and the Higgs boson. The Genome Project is small potatoes compared to these truly Big Science projects.

Ray White believes that the Genome Project is moving biology and biologists into a new willingness to invest in tool-making, something that is common in Big Science projects like the Hubble Space Telescope or the Superconducting Supercollider. The new tools for the Genome Project will include automated DNA sequencers and computerized databases for DNA sequence information. The Genome Project, White feels, will have a similar impact on all of the biological sciences. In the past the development of tools and instruments for biology was something that happened piecemeal. A researcher would invent something he or she needed for a specific project. Researchers could make an investment in technology, but it was not a large one. The technology development rarely went beyond the one laboratory. The amount of money involved was only as large as could be justified by the hypothesis behind their experiments. That, of course, limited the scale of technology that could be developed. It virtually eliminated the possibility of any major development of new scientific instrumentation. The result, White has said, is that biology has been at a "cottage industry" level. A cottage industry level for molecular biology will simply not support the effort needed for a project of this size.

Ray White frowns on what he calls "the bugaboo of Big Science." That is the concept of mega-instruments with hundreds of investigators, with researchers having to sign up months in advance to use the mega-instrument for a few hours or days. White does not see the Genome Project as ever being Big Science in that sense. The Genome Project is Big Science for biology, but it is not Big Science in the way the Superconducting Supercollider is Big Science for physics. It might be more accurate to call it "small Big Science," or "big Small Science."

One reason is that the National Institutes of Health is for the first time consciously investing in tools and instrumentation to make the work of researchers more effective. In the context of the NIH's budget, the proportion going to the Genome Project is actually rather small. It is, however, a small percentage investment in basic technology in order to do the science. Large corporations like DuPont and Hoffmann-La-Roche have done this for years. It is a relatively new development for NIH, and especially for the field of molecular biology. One consequence of this will be changes in the overall infrastructure of biology, to allow room for large projects that will focus on the development of scientific instrumentation.

Another difference between the Big Science of the Genome Project and other Big Science projects is the distribution of the work. The Genome Project is being much more widely distributed than other Big Science programs in other disciplines. There are not simply two or three huge laboratories of scientific installations doing all the work. Instead, the Genome Project is being carried out through a distributed network of installations. The computing power will be distributed in the same way. Small-scale high-tech instruments will be distributed to the laboratories that want to use them. In fact, this is really nothing more than traditional biological Small Science being carried out by small groups. The challenge is going to be finding ways to retain as much of that creativity as possible, in the context of an infrastructure that will be—must be—larger than anything biology has had before.

DOE VS. NIH

The unease within the biology community over the "Big Science" aspect of the Genome Project was most clearly embodied in the jurisdictional battle that took place between two giant federal bureaucracies—the U.S. Department of Energy and the National Institutes of Health. The formal proposal of the Genome Project had come from the DOE, in the person of Charles DeLisi of the OHER. The Energy Department had many years of expertise in successfully running programs of great size. It had a wealth of scientific expertise in genetics at several of its national labs, including Los Alamos, Lawrence Berkeley, and Livermore. The NIH, on the other hand, was an agency into which the biological and medical aspects of the Genome Project fit quite well. It, too, had on tap many eminent scientists and researchers. It did not have much skill or history at managing "Big Science" projects, but it wanted very much to have a major say in the Genome Project. The result was the development of an intense interservice rivalry between the two agencies. This was accompanied by the establishment of parallel committees in the two agencies and a massive lobbying effort in Congress, from whom would come the funding for any Genome Project.

Nevertheless, the two parties were talking with each other. A series of workshops involving people from both agencies resulted in a general agreement on which direction to take. The consensus was to begin with the creation of a physical cosmid map. This would be an ordered set of DNA fragments, each about 40,000 base pairs long. The next step would be to locate genes on the cosmid map. Concurrently with this, of course, people like Ray White would continue to create genetic linkage maps using RFLP markers. The important point was that any massive sequencing effort for the entire 3 billion base pairs of the human genome would be put on hold. Walter Gilbert's proposal lost out. Both agencies instead came to agree with Lee Hood of the California Institute of Technology, who advocated holding off on massive sequencing projects until automated sequencing technology was advanced enough to make it a relatively cheaper and faster enterprise.

Despite agreement on this, other questions remained unresolved though the middle of 1987: Should the Genome Project be a large, centralized effort? Or, should it be carried out as "Small Science" in many labs around the country, with some agency as the overall coordinator? Or, should it be an effort carried out in large regional centers? Who should be the lead agency, DOE or NIH? What should the focus be: strictly the human genome, or ought other complex genomes be included? Should there be a scientific advisory committee? Which agency should be home to such a committee, if it is established?

At this point there was no formal interagency coordination. Indeed, it seemed that both the DOE and the NIH liked it that way. They were happy to talk with each other, but fully intended to go their own ways. The DOE, with DeLisi and David Smith in the lead, began planning to pursue a very aggressive course, albeit somewhat more diplomatic than it had been in 1986. It targeted $12 million in 1988 for genome-related work. An April 1987 report from the department's Health and Environmental Research Advisory Committee (HERAC) called for an even more ambitious program, recommending budget increases to eventually $200 million a year. The HERAC report plainly stated that "DOE can and should organize and administer this initiative." David Smith, however, responded to the report by calling its goals "too ambitious." The HERAC report, he said, seemed to embrace all of genetics in its vision of the genome initiative, and that was too broad for the DOE. That agency excels at developing technological tools, he said, and that's probably what we will focus on. Actually using the tools and technology to map and sequence the genome, he added, might be better left to "the medical biological community, and that's in Bethesda," the home of the NIH.

Conversely, the NIH at that time seemed to be acting rather reluctantly, almost as if it were afraid to tackle a project as large as the Genome Project. One reason for that public stance, it seemed, was that NIH was *already* heavily involved in gene-mapping. In 1987 it was spending over $300 million on such research, with a third of that on the human genome alone. Rachel Levinson, at that time working in the office of then-NIH director James Wyngaarten, suggested that the amount of money that the NIH was already putting into genome-related research clearly proved the agency's engagement with the Genome Project, and its intention to increase that commitment. In other words, the Genome Project was nothing new: the NIH was already doing it.

Others disagreed strongly. Smith of OHER noted that most of the $300 million was going to work not connected with the development of a physical map of the genome. The DOE's efforts, on the other hand, were going to be tightly focused on just such goals. We think this is something new and different, Smith and others at DOE seemed to be saying; and we plan to be the leaders.

Most damning, perhaps, were similar comments from one of NIH's friends, Lee Hood of the California Institute of Technology. To Hood, the statements of Levinson and others at NIH revealed that agency's support for the Genome Project as lukewarm at best.

In point of fact, the NIH *was* lukewarm about the Genome Project as envisioned by the DOE and its supporters. Part of its reluctance to get aboard the Genome Train was the agency's traditional commitment to research initially proposed by individual scientists—classic "Small Science" biology, in other words. The Genome Project would be biology's Big Science effort, and the NIH had no expertise at such an undertaking. Part of it had to do with the large amount of technology development, which also fell outside NIH's sphere of competence. The agency typically supported science, not tool-making. Still another aspect of the agency's reluctance—and perhaps the major one—was the fear that a Genome Project would drain money from other areas. Wyngaarten and others wanted to be assured that such a massive initiative would not rob money from other NIH programs.

And in spite of its vacillations, NIH did not want to relinquish control of a Genome Initiative to the DOE. The agency felt strongly that any such project involving biology belonged in its bailiwick. It was a curious and rather frustrating situation for many people. Hood, for example, remarked at the time that DOE had the Big Science expertise, but NIH had the biology skill.

The White House's Biotechnology Science Coordinating Committee (BSCC), the National Science Foundation (NSF), the Congressional Office of Technology Assessment (OTA), and the DOE all had committees studying the matter of who should lead the Genome Project. There was even some talk of not needing any lead agency at all, as long as the DOE and the NIH continued talking with one another, and began making some effort to coordinate their work. At this point, midway through 1987, in stepped James Watson.

Watson made two points clear: (1) Bureaucrats must not be in charge of coordinating Genome Project work done by the agencies involved. The coordination must be done by scientists. (2) It would not be possible to carry out such a major scientific and technical project without a lead agency. There's only one genome, Watson said, and we need one lead agency.

He proceeded to plan a meeting that fall of people seated on all the currently existing committees looking at the Genome Project. That included Victor McKusick of Johns Hopkins, Lee Hood of the California Institute of Technology, Frank Ruddle of Yale, and other major figures in biology and genetics. The meeting's purpose would be to establish precisely the kind of scientific advisory group Watson was talking about. And one of the questions the group would answer would be: Who will be the lead agency?

At the same time, pressure began building from another quarter for a resolution of the internecine rivalry. Congress was beginning to take serious notice of all the talk about a Genome Project. The concept got hooked into the political and economic concern over U.S. technological competitiveness and superiority. Several senators and congressmen made it clear that they wanted to give someone plenty of money to map the genome. It also started becoming clear that this would be "new" money, not stolen from other scientific or technical programs. David Baltimore, for example, would not have to worry about funds first earmarked for AIDS research being shifted to genetic mapping. The battle over adequate AIDS research funding, so deftly chronicled in Randy Shilts' book *And the Band Played On*, would continue. But the Genome Project would not become a potential adversary in that desperate fight.

As the lobbying and maneuvering between DOE and NIH continued through 1987 and into 1988, still another message began to be heard from Congress: You folks had better get your act together and decide who's boss, or *we* will decide for you. Several bills were introduced in Congress to appoint one agency or another to lead the Project. As 1987 became 1988, the NIH's position began to change. Their lobbying became stronger, more pointed. The statements from the agency began to sound more aggressive. The NIH began to accept the idea that the Genome Project was indeed something new, some-thing qualitatively different from what they had been funding so far. In February director James Wyngaarten announced that he would establish an Office of Human Genome Research within the NIH. And in May the word began to leak out: Wyngaarten had asked James Watson to head the office.

Perhaps this was part of a deliberate strategy to stop the DOE's drive to head the Genome Project. Perhaps not. But the effect of the announcement was to place the NIH clearly in the ascendancy. DOE and the NIH would eventually get to-gether and thrash out a Memorandum of Understanding about coordination of their genome mapping programs. But when Watson agreed to accept the position of head of the Office of Human Genome Research (which he would hold while con-tinuing to run the Cold Spring Harbor Lab), the question of which was the *de facto* lead agency for the Genome Project was effectively settled. To the surprise of many, it is the National Institutes of Health.

THE BUREAUCRATIC BOGS

One obstacle that the proponents of the Genome Project wor-ried about from the beginning was the immense volume of data that would come pouring out of the Project. Efforts to upgrade computer database capability were planned nearly from the beginning. However, the Genome Project boosters did not initially anticipate a problem typical of any huge scheme. The "Big Biology" effort of the Genome Project is going to produce a great deal of bureaucratic red tape.

By the summer of 1989 it was clear that even within the United States, the Genome Project was garnering increasing interest from Congress. The proclivity of the legislative branch to try to "micro-manage" many programs being carried out by executive-branch agencies and departments—the Depart-ment of Energy and the National Institutes of Health are two—was well known. The international nature of the Project insured the interest of other governments, as well. The attitude of many Genome Project researchers was summed up by Norton Zinder of Rockefeller University, chair of the program advisory committee for the NIH's Genome Initiative programs. "If the governments will leave us alone," Zinder was quoted as saying in *Science* magazine, "we will be all right. But if we are forced to act as delegates of our countries, rather than as scientists engaged in an exciting international project, then it won't work."

It will be difficult to avoid the Bureaucratic Bogs, however. International cooperation will be essential to the success of the Genome Project. Each country that has researchers working on the Project will see its effort as contributing to national pride and prestige. Even countries like Italy, with at best a modest capability in the biological sciences, have started their own national genome projects. More importantly, it will be necessary to coordinate the pooling and distribution of data produced at different laboratories in various countries. The establishment of several regional database locations will help. It will also inevitably result in still another layer of bureaucracy for the Genome Project, with a concomitant increase in the volume of paper being pushed from desk to desk, file cabinet to file cabinet. Not even the extensive computerization of the Genome Project will eliminate that.

As a result, one increasingly necessary task of Genome Project advisory committees like that headed by Zinder will be organizational management. Finding ways to keep bureaucratic red tape to a minimum and scientific communication to a maximum will become more and more significant to the success of the Genome Project.

BIG BIOLOGY AND ANTITRUST LAWS

Another problem for the Big Biology being created by the Genome Project might well be legal. George Poste is the head of research and development for the Smith Kline & French drug company. At a 1989 medical conference he warned of the dangers of antitrust action against drug companies involved in the Genome Project. Poste believes that it will not be possible for any single pharmaceutical corporation to solve all the technological problems associated with the Genome Project. It would simply cost too much. There is thus no chance, he claims, that any one drug company will be technologically self-sufficient and able to monopolize the medical market that will be created by the mapping of the genome.

Instead, Poste thinks different companies will soon band together in several technical consortia, working together to solve the technological problems and share the costs. That strategy, he thinks, could bring the companies—and the Genome Project—into conflict with U.S. antitrust laws. Big Physics doesn't have a problem with antitrust laws. That science's huge projects, like the Superconducting Supercollider, are funded primarily by the federal government itself, and carried out by universities. The same is true of Big Astronomy projects like the Hubble Space Telescope or SETI, the Search for Extraterrestrial Life. The defense industry's various consortia and collaborations, which are closer in structure to the Genome Project, are ignored by the government's antitrust lawyers because they are considered to be "in the interest of national security."

But is the Genome Project? Poste believes that various governments around the world do not understand that companies will have to work together to turn the scientific knowledge produced by the Genome Project into practical medical technology. Neither does the general public, including citizen activist groups suspicious of huge pharmaceutical companies. Anti-science and anti-technology activists like Jeremy Rifkin might also seize upon antitrust lawsuits as a way to stop the creation of pharmaceutical corsortia, and thus slow down or stop the Genome Project. No matter what happens, says Poste, the public debate over the Genome Project is likely to become rather fierce in the next several years. Everyone has a personal stake in it, because this first Big Biology project literally touches each person's health and life.

THE BIGGEST CONCERN

In the end, what probably worries researchers like David Baltimore the most is not so much the money or size of a Big Science Genome Project, but the consequences such an endeavor might have to *the way biology is done as a science*. And in this, Baltimore and other critics are correct. The Genome Project will forever change the way biology and genetics are done. That does not mean that all biology and genetics will soon be carried out in massive, government-sponsored labs, or that all biomedical science will be driven by political clout and huge federal grants. That scenario is not likely to come true, even in the case of the Genome Project itself. Most of the Genome Project leaders have taken steps to assure that the research and technological development for the Project will continue to be carried out in many laboratories in the United States. The growing international flavor of the Genome Project also assures the wide distribution of the mapping and sequencing efforts involved.

On the other hand, the DOE's national laboratories at Berkeley, Livermore, and Los Alamos are now firmly entrenched as major centers of biological science. Indeed, they were before the inception of the Genome Project, but with its rise to prominence they are now more firmly committed than ever to playing major roles in the genetics and biology of the future. Other laboratories are also rising to prominence—Ray White's at the University of Utah; Maynard Olson's in St. Louis; Lee Hood's at the California Institute of Technology in Pasadena, California. Companies like Collaborative Research are also playing increasingly important roles in the Genome Project, and corporations like Genentech and Biogen have already come to the forefront of genetic engineering and biomedical research.

What the Genome Project is doing, and will do, to genetics and biomedicine as scientific disciplines, is change their style and magnitude. The Genome Project is introducing biologists to computers, for example. That in itself will forever alter the way biology is done. The next generation of biologists will use desktop—even laptop—computers as lab notebooks. The old paper lab notebook is on its way out. The high-powered computerized databases that the Genome Project will need to store and correlate data will also have a profound effect on biology and genetics. New linkage analysis programs and protein sequence algorithms will speed up the process of mapping and sequencing DNA and proteins. That in turn will revolutionize the creation of new drugs and medicinals.

Biologists, like other scientists, have always shared information informally with others, even competitors. To reach its goals, the Genome Project will have to rely on that sharing to a much larger extent. The CEPH collaboration spearheaded by Jean Dausset is a good example of the future of data-sharing in biology and genetics. Today, many research papers in nuclear physics have dozens of coauthors. Biology reports published in the professional journals may have half a dozen. In years to come, though, more and more biology papers will have as many coauthors as do physics reports.

In these and other ways, biology and genetics will become small Big Sciences. The Genome Project is a major contributor to that change.

And there is nothing anyone can do about it.

6

"Owning" the International Genome

> . . . . my physicians by their love are grown
> Cosmographers, and I their map,
> who lie flat on this bed.
>
> JOHN DONNE
> "Hymn to God, My God, in My Sickness"

W ALTER GILBERT is good at many things. Three of them are doing science, creating companies, and getting people riled up. In 1980 he won the Nobel Prize for the first. Not long afterward he did the second by helping start Biogen, Inc., one of the world's first biotechnology companies. At the 1986 Cold Spring Harbor meeting he got people's attention by doing the last.

WALTER GILBERT AND THE GENOME CORPORATION

In early 1987 Gilbert combined all three. He announced that he was going to start a new company, called Genome Corporation, and decipher the human genetic code. His partner in this new venture was Jeffrey Wales, the chief executive officer of the Polygen Corporation, based in Waltham, Massachusetts. What's more, Gilbert added, he planned to *copyright* the sequences he came up with and sell the information at whatever price the market would bear. The whole concept was more than enough to ruffle a lot of scientific feathers.

Just a year earlier, Gilbert had emphasized that the Genome Project would cost $3 billion and take 15 years. Now, though, he was claiming that he and his new company would be able to sequence the genome in just ten years, for only a tenth of his earlier estimate. The key, said Gilbert, was automating the sequencing process. Gilbert was anticipating the swift development of highly computerized and automated DNA sequencers. He hoped to be sequencing up to a million base pairs a day. At that rate the Genome Corporation would indeed sequence all three billion base pairs of the human genome within a decade.

Besides the technological leap into super-automated sequencing, Gilbert was also proposing a major commercial push. As he and his partner Wales travelled about the country courting startup money, Gilbert told potential investors that his proposed company could market small portions of the genome within three years of startup.

If that sounded a bit optimistic to some people in 1987, his proposal to copyright the genome in his name was downright outrageous. Many genetic researchers, not to mention lawyers, felt the genome is the ultimate case of information in the public domain. No one, they claimed, can copyright something that is the essential inheritance of the human species. Since the genome contains all the information needed to create a human being, copyrighting the genome would be a biotech version of slavery. Gilbert, in response, pointed out that we live in a society in which people do own other living things such as pets, livestock, and plants. Further, he said, the human genome is not itself a human person. So ethical objections to copyrighting the genome do not hold water. Besides, it was not so much the genome itself that he wanted to copyright, he said, but the physical sequence of representative letters—the As, Ts, Cs and Gs—that is written down as the sequence. That was what he would copyright, just as a writer copyrights the sequences of letters and words that make up a novel, or a musician copyrights the sequence of written notes that make up a song.

CHANGES IN ATTITUDE

Gilbert's proposals for marketing the genome were very different from the way things were normally done in biology. However, changes in attitude and behavior were already taking place. Researchers, for example, had long been open in their willingness to share DNA clones with their colleagues. For example, a group of scientists at the University of Oregon Health Sciences Center in Portland, might have a DNA probe for the gene coding for met-enkephalin (one of the endorphins). Researchers at Columbia University might need clones

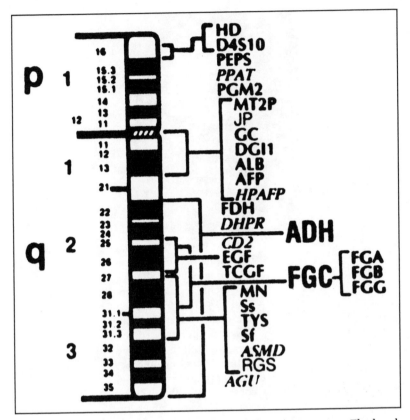

This is a pictorial representation of a human gene map. The banding patterns on the chromosome are produced in the laboratory with special staining techniques. Each chromosome has two arms, a short or "p" arm and a long "q" arm. The different divisions in each arm are numbered, with "1" starting at the central division point or centromere and the largest number just before the outer end or "telomere" of each arm. The rough location of different genes or genetic markers is indicated by the abbreviations. For example, on chromosome 4 the abbreviations "HD" and D4S10" at the outer end of the p arm stand for genetic markers associated with Huntington's disease.

of that probe for an experiment. Even if both teams were working on the same problem, the Columbia people would not hesitate to call the University of Oregon team and ask for some clones of the probe. The people in Portland would not

Nobelist Walter Gilbert has been at the forefront of genetic engineering science and technology for more than 20 years. He has been one of the earliest enthusiastic proponents of the Genome Project, but some of his suggestions and actions have also engendered controversy.—Photo courtesy of Harvard University.

hesitate to send them. "Clone by phone" was a common practice. In July 1987, however, C. Thomas Caskey of Baylor College would be quoted in *Science* magazine as saying that he was already detecting a "tightening of this attitude" when he called fellow researchers to ask for copies of DNA clones.

Genomic research was beginning to become more "proprietary." Some scientists were beginning to think like Gilbert, that information from genomic mapping and sequencing could be *owned*, and therefore bought and sold. The patenting of genetically engineered microbes had already taken place in the biotechnology field, and computer programmers were beginning to try to copyright their software codes. Now geneticists and biologists were beginning to think about the same thing. Companies like Collaborative Research, Incorporated (CRI) were founded on the premise that they could commercialize genetic linkage and physical maps. Researchers were beginning to realize that there might be money—big money—in discovering new gene loci. Especially if the loci were for genes connected to serious illnesses.

In June 1987 the Howard Hughes Medical Institute (HHMI) and the Congressional Office of Technology Assessment (OTA) sponsored a one-day workshop in Washington, D.C. on "Issues of Collaboration for Human Genome Projects." HHMI was already spending $1.5 million a year on genome mapping and sequencing projects. (Ray White's lab at the University of Utah was running one of them.) The conference included lawyers, ethicists, representatives from industry, public and nonprofit agencies, and genetic researchers. The consensus was that problems of collaboration could easily become more difficult as the Genome Project increased in size and scope. It would be necessary to set up methods in advance to ensure the open exchange of information and materials. In one respect, this was not so much a symptom of the Genome Project as Big Science as it was a long-standing situation in biology. Some people have always been willing to share information, noted Leroy Hood (one of the majority of researchers who shares), and some have not. That had not changed. What had changed was the open speculation about copyrighting or patenting information.

COPYRIGHTING AND PATENTING

Not surprisingly, many people at the OTA workshop expressed concern about the ethics of trying to copyright the genome sequences. Frank Ruddle of Yale University commented that such a move would make him "very uncomfortable." Genomic information is so important, he said, that he didn't see how it could be considered proprietary, something that could be owned, copyrighted, or patented by either an individual or a company. Ruddle agreed that genomic information was monetarily valuable. That information can and will be sold. Indeed, he felt that such value worked as a goad to many scientists to carry out the often-tedious work of mapping and sequencing segments of the human genome. The problem, he said, was that not everyone would have access to the information were it available on a strictly "cash and carry" basis. Such an arrangement might well be possible, even likely, given the current state of affairs. But was it ethical? George Cahill, who oversees the Institute's programs related to the Genome Project, would remark on the importance of looking at "the bucks to ethics ratio."

One cogent observation came from Robert Cook-Deegan, who was in the middle of authoring a major report by the OTA on the Genome Project. Cook-Deegan noted that the Genome Project would eventually produce a body of information that could be considered a national resource (if not a global resource) and that should be available to all. In this respect the Project would differ from all other projects or experiments ever attempted by either the community of biological scientists or by for-profit corporations. The Genome Project, in other words, is *not* "business as usual." That in turn implied that sharing data for the Genome Project was of the utmost importance. Researchers would have to talk with each other, even if the talking involved so-called proprietary data.

In June 1987 no such procedures existed. DOE and the NIH were pursuing their own separate genome initiatives, and only just beginning to talk with each other about collaborations and sharing of data. Although the Genome Project did not yet officially exist, it was soon to move into the arena of Big Time Politics.

Again and again, conversation at the OTA workshop came back to Walter Gilbert's plans to copyright human genome data. Not even the lawyers at the workshop were sure about whether or not such a thing was legally possible. One view, not widely held but still voiced, was that DNA sequences *can* be copyrighted. This view held that a genomic sequence is like a computer code. Computer codes can be copyrighted, and therefore so can the DNA code. Gilbert's position was similar to this. Some of the workshop participants reluctantly ceded this point. Others did not.

Some, like George Cahill of HHMI, felt that *something* was copyrightable but not the DNA sequence itself. Rather, said Cahill, it was possible that Gilbert and others could copyright the *format* in which the sequence is presented. This position presaged the huge court battles that would be waged in later years by computer corporations over the "look and feel" of certain computer programs. If a computer company released a program that presented information in a series of "windows" on the computer screen—as one such company did—could that company claim an exclusive copyright to that format? If so, would companies that developed other programs using windows have to pay the first company royalties? In somewhat the same way, Cahill was saying that a biotech company could copyright the format in which DNA sequence data was presented to a user. However, the company could not copyright the information itself.

The bottom line for Cahill, however, was not the issue of copyrights, but what copyrighting DNA data would mean for the rapid exchange of genomic information. DNA sequence information developed by Gilbert's Genome Corp.—or anyone else—would be valuable only as long as it was not publicly available. Then its value to the company would disappear. The company could not make money selling the information when anyone could get it for free. That in turn suggested that it would be economically worthwhile to restrict access to the information for as long as possible.

And that, Cahill charged, was totally antithetical to the very basis of science itself. If scientific information does not flow freely and is not quickly available to anyone, then good science cannot be done. Science as a process of understanding the cosmos can flourish only in an atmosphere of openness to information and observation. When that is lost, science dies.

COLLABORATIVE RESEARCH AND RAY WHITE

The concern over withholding genetic information was not a theoretical one. At the time the OTA workshop was held, a controversy was already raging over the practices of Collaborative Research, Inc. (CRI), which was developing RFLP marker maps for several human chromosomes. By mid-1987 CRI had spend more than $10 million and developed about 600 markers. Its avowed goal was to become the leading company the field of diagnostic tests for cancer and genetic diseases. CRI, like other genetic mappers, planned to use RFLPs to map the rough locations of genes that were linked to diseases like cystic fibrosis. Ray White of the University of Utah was doing the same thing. However, White was depositing his probes and markers with the American Type Culture Collection, where they would be available to anyone. CRI was not. Scientific collaborators had access to CRI's probes only if they signed confidentiality agreements.

The tension between CRI and others (including White) arose because of the way diagnostic tests are developed in response to genetic linkage maps. It is not enough to simply find a marker that is linked to a gene on some chromosome. The first markers may be relatively far from the gene itself. For an accurate genetic screening test, markers must be very close to the gene, very tightly linked to it. If investigator A announces that he or she has found markers linked to a gene in a particular location on a chromosome, investigators B, C and D will quickly try out their suite of DNA probes for that chromosome. They will want to find a marker that is more tightly linked to the gene than is the one found by investigator A. In terms of keeping the potential competition at bay, therefore, it makes sense to keep one's mouth shut about one's newly-discovered markers.

It is very likely, almost a certainty, that such behavior has already been taking place. In 1987 a series of papers were published almost simultaneously which announced a possible location for a gene for Alzheimer's disease. One participant at the OTA workshop noted drolly that it was unlikely everyone discovered the gene at exactly the same time. The truth was, as another participant, a lawyer, noted: "Researchers sit on stuff all the time, until they are sure." A patent attorney for Hoffmann-La Roche told the workshop participants that biotech companies actually make information available to the public much more quickly than patent attorneys would like. Most patent attorneys would apparently prefer their researchers to withhold release of their scientific findings for at least a year and a half. That would make it easier for the lawyers to get patents in other countries. For many scientists, however, the impulse remains to publish their results as early as possible, in order to establish scientific priority. That might not be important if all one has found is a more accurate location for some obscure gene. But if it's the location of the gene for cystic fibrosis, or Alzheimer's disease, or Huntington's disease, scientific priority could translate into a Nobel Prize. And despite any protestations to the contrary, every scientist would love to win it.

When Gilbert first made his announcement, many observers thought he would have little trouble raising the money. Orrie Friedman, chairman of Collaborative Research, Incorporated, which was itself in the genome-mapping business, was quoted as saying that Gilbert "was in a class by himself." On the other hand, those who knew Gilbert as a manager of a large corporation were not that impressed. Robert A. Fildes had been Gilbert's second in command at Biogen before he left, later to become president of the Cetus Corporation in California, now one of world's most prominent biotech companies. Fildes expressed the opinion that Gilbert, though a brilliant scientist, had never shown any great business sense.

Others must have felt that way. Despite all his efforts, Walter Gilbert never quite got the Genome Corporation off the ground. It did not get enough startup funds. By the time the Genome Project had taken off in 1989, the Genome Corporation was nothing more than a company on paper.

Collaborative Research was not. However, just what the company will eventually produce was still somewhat fuzzy at the end of 1989. In 1987 CRI made two major claims to fame. One was the discovery of the location of the gene for cystic fibrosis. The other was the announcement of the "first" genetic linkage, or RFLP, map for the entire human genome. Both fell short of reality.

Ray White was one of the early supporters of CRI's work. He helped as a consultant. However, White and the company soon came to a parting of the ways. There have been few kind words between them since. White had advised them early on to avoid spending company money on the development of a RFLP map. His suggestion was that they go to some non-profit foundation or group and get the work publicly funded. However, the CRI management decided otherwise. The company made a $10 million to $12 million dollar investment. It is not obvious to White how they will be able to recover that commercially. One way might be to try to sell or license their DNA probes to other researchers.

Ironically it was largely the enthusiasm of David Botstein for the RFLP mapping project that eventually convinced CRI to put money in it. Botstein, of course, had been White's collaborator on the seminal 1980 RFLP paper. However, he apparently felt that White was not moving fast enough on the development of a RFLP map, and that Collaborative Research should push ahead with it. It may nor may not be significant that Botstein left CRI in 1988, moving to Genentech Corporation near San Francisco as a vice president.

THE RFLP MAP CONTROVERSY

Collaborative Research's release in 1987 of a "complete RFLP map" of the human genome led to what Ray White himself would call "one of my most indiscreet moments." White and his team, of course, were working slowly but steadily on the creation of a ten-centimorgan resolution genetic linkage map of the human genome. They knew that CRI's researchers were, too. The lead scientist at CRI on the project was Helen Donis-Keller, a brilliant scientist who is now working at Washington University in St. Louis. White and Donis-Keller know each other well, and each has great admiration and respect for the other's work. That appears still to be the case, despite White's episode of "foot in mouth disease."

White did not think the company would take what he considered the "premature and dramatic step" of announcing the completion of such a map in mid-1987. He suspects that Donis-Keller was also somewhat surprised. White's scenario is that the top management of CRI was afraid that he, White, was about to announce the completion of such a map. They knew that White's team had many more RFLP markers mapped onto the chromosomes of three times as many people. Collaborative Research had assumed (White thinks) that they would be the people who would make the genetic linkage map. Now it began to dawn on them that this might not be the case. They would be *a* major player, but *not the only one*. They decided to preempt White, and go public first. The result, said White, was the publication of a RFLP map of the human genome containing huge gaps in its coverage. For example, the Collaborative Research map of chromosome 10 was piecemeal at best. There were no markers at all on the short arm of chromosome 10, and only a third of the long arm had good coverage.

White admits that he overreacted to the announcement of Collaborative Research's RFLP map. The first he heard of it was from a reporter from the *Wall Street Journal*, who called him on the phone asking for quotes for an article about the map. White was, he admitted, "at the top of my adrenals," and the *Journal* reporter got the full benefit of White's ire. The quotes (everything White told the reporter was "on the record" because he had not specified otherwise) in the resulting article were angry, bitter, and hurtful, particularly to Helen Donis-Keller. White quickly regretted what he had said, and personally apologized to Donis-Keller.

Nevertheless, White remains convinced that Collaborative Research prematurely released an incomplete genetic linkage map of the human genome. He feels that it was done partly out of the need of the company to assert that they had a major role in gene mapping. This was combined, he thinks, with the need of the company to try to establish their hegemony in the business world. Potential profits from diagnostic tests and genetic engineering techniques resulting from the map were not far from the minds of the company's leaders.

Despite all the controversy, however, White also thinks the Collaborative Research RFLP map has considerable usefulness. This is particularly so if it is treated as a preliminary but good sketch for a more complete genetic linkage map. Meanwhile, White and his team at the Howard Hughes Medical Institute labs in Salt Lake City are continuing their work on a ten-centimorgan RFLP map of the genome. Their maps were to be released through 1990, one chromosome at a time. They have chosen not to release any map until they feel they have enough information to make it complete. Their first ten-centimorgan map will be of chromosome 10. Closure for data submission was Dec. 1, 1988. The final map of chromosome 10 was put together in early 1989. Others were to follow over the next year or so. Sometime in 1990 or 1991, felt many at that time, a truly complete genetic linkage map of the human genome would finally exist, with a resolution of ten centimorgans.

THE OTA REPORT

Even as researchers like Donis-Keller and White were racing to create the first genomic maps, a government agency was finishing a report on the "potential" significance of the Genome Project. In April 1988 the Office of Technology Assessment released its report on the Genome Project, *Mapping Our Genes—The Genome Projects: How Big, How Fast?* Project director for the report was Robert Cook-Deegan. On the OTA's advisory panel for the report were several people who would continue to be major players in the mapping effort, including George Cahill of the Howard Hughes Medical Institute (HHMI), Nancy Wexler of Columbia University, Ray White of HHMI and the University of Utah, Walter Goad of the Los Alamos National Laboratory and former head of that lab's GenBank database, and Leroy Hood. The 218-page document was the third major government report in a year on the Genome Project. The others were a report from the Department of Energy released in April 1987, and one from the National Research Council (NRC) in February 1988. The OTA report was by far the most sophisticated of the three, and benefited from the summaries and comments of the earlier two.

Mapping Our Genes also focused on the political and organizational options the Project could take to be successful. The timing was good. Several House and Senate committees and subcommittees were considering bills that would establish various government boards and commissions to oversee biotechnology development. One House committee held hearings in April on the scientific and technological needs of a genome project. The NIH had also put together an Ad Hoc Advisory Committee on Complex Genomes. That was part of the Institute's effort to capture the "Genome Project high ground" from the Department of Energy. The 1987 DOE report, not surprisingly, had recommended that that agency was best suited to be the leader of the Genome Project. There was no doubt that DOE was experienced at managing large, complex technological projects. However, many scientists felt that the NIH was the better agency to lead the effort, since human genetics naturally fit into the NIH's sphere of endeavor. The NRC's report in February had recommended that one agency lead the Genome Project, but made no specific recommendations on which one.

The OTA report did, however. It recommended that NIH be the lead agency for the enterprise. It also suggested that, among other things, an interagency task force be established to coordinate the activities of all the government agencies involved, including DOE and NIH. Within a year, the OTA's recommendations became reality. DOE's Charles DeLisi, that agency's mover and shaker in the Genome Project, later left to become a professor at the Mount Sinai School of Medicine in New York. His boss at DOE, David Smith, continued to push DOE's role in the Project. However, NIH's aggressive lobbying for leadership soon put it on a par with DOE. That agency continues to play a major role, of course. Its national laboratories at Los Alamos, New Mexico, and Berkeley and Livermore, California have major Genome Project research teams with people like Tom Marr and George Bell (Los Alamos), Charles Cantor (Berkeley) and Anthony Carrano (Lawrence Livermore).

The OTA report made several other cogent points that later became part and parcel of the actual Genome Project. One was that "there is no single human genome project, but instead many projects." Other species with complex genomes would also have to be included, including the nematode, the mouse, and *E. coli*. Another point had to do with money. The OTA report estimated that the first year's funding of the Genome Project would probably run about $47 million. That was in fact close to the mark. The report also offered funding estimates for the following four years. However, the OTA refused to go beyond a five-year projection. The NRC report had projected costs over an estimated 15-year lifetime for the Genome Project.

THE INTERNATIONAL GENOME

Researchers from countries around the world are involved in the Genome Project. Indeed, the existence of MIM shows that

Charles Cantor, director of the Human Genome Center at the Lawrence Berkeley Laboratory in California, speaking to the Human Genome Center Advisory Committee.—Photo Courtesy of Lawrence Berkeley Laboratory.

genetic mapping has been going on for more than seven decades now, and its pace has been increasing exponentially in the last few years. The establishment of a formal "Genome Project" is just the latest episode in a scientific drama that reaches back to the beginning of this century.

International interest in this formal Genome Project is increasing. At first, though, the American initiative met with less than extreme global enthusiasm. The only notable exception was Japan, which was intensely interested in the Genome Project nearly from its inception. The shift in interest from sequencing to mapping the genome was also international. At the July 1986 NIH meeting in Bethesda, John Tooze of the European Molecular Biology Organization (EMBO) noted that he and his colleagues were not all that interested in sequencing the genome until the process could be more automated. British researcher Sydney Brenner was quoted as saying that "the idea of trudging through the genome, sequence by sequence," did not have much support in Great Britain. Japanese researchers, by contrast, were telling people that they were going to develop super-computerized machines that would be sequencing a million bases a day by 1988. They would, they said, then proceed to sequence chromosome 21 in just five months. Chromosome 21 is the smallest human chromosome, with about 48 million bases. At the same time, Japanese researchers also said they had no plans to go ahead with a full Big Science sequencing project. The organizational structure for such an undertaking, they said, did not yet exist in Japanese science.

Within two years, though, interest in the Genome Project had spread worldwide, both in the scientific community and the political arena. National governments saw the Genome Project as a doorway to international prestige. Headlines like "Italian Scientists Map Chromosome—" must have flashed through politicians' imaginations. It sounded good, and grand, and very patriotic. The money started flowing. More important than the national commitments, however, were the ongoing international collaborations. As the Genome Project entered the last decade of the twentieth century, it had become truly international in scope. That was signaled in a significant way by the formation in 1988 of an international coordinating organization.

And who was there at the beginning of its creation? None other than Victor McKusick.

HUGO

One of the most important developments in the international scope of the Genome Project has been the founding of HUGO, the Human Genome Organization. According to McKusick, HUGO was conceived on April 30, 1988. It happened during that year's genome mapping symposium at Cold Spring Harbor. The idea of an international organization to oversee and encourage genome mapping research had actually been tossed about by researchers for over a year. During a small gathering at the symposium, the idea took shape. The Genome Project was soon to become a reality. The time for talk was over. Sydney Brenner suggested the name and it caught on.

McKusick and a few others began contacting various genetics researchers and putting together a list of people who were willing to serve on a founding board of directors and advisors. After a five-month formal gestation period, HUGO was born on September 6 and 7, 1988, at a meeting of its founding council in Montreaux, Switzerland. The organization was later incorporated in that country. Its membership is elected, and it has officers and an executive committee to direct its operations.

HUGO is not really a scientific association or society, like the American Association for the Advancement or Science, or the American Physical Society. Rather, it is more of a "U.N. for the human genome," as HUGO member Norton Zinder once put it. It is similar in structure and purpose to EMBO. Indeed, McKusick and the other founders consciously modeled it on that well-known European scientific organization. Like EMBO, HUGO was started with private funding from the Howard Hughes Medical Institute, the Imperial Cancer Research Fund or ICRF (a private medical charity in Great Britain), and some private Japanese sources.

HUGO has four main goals or functions:

■ To help coordinate research on the human genome and to provide international training programs on the relevant methodology;

■ To arrange the exchange of data, samples, and technology relevant to genome research;

■ To foster parallel studies in model organisms such as the mouse and coordinate that research with the human Genome Project;

■ To provide public debate and develop guidelines on ethical, social, legal, and commercial implications of the genome project.

Implementing these four goals will not be easy. HUGO is young, and still has relatively little funding. Nevertheless, its founders are optimistic and organized. In order to implement the goals, they hope to carry out a variety of activities:

■ Planning regional centers for large-scale mapping and sequencing, which would also coordinate major resources such as databases, collections of DNA clones, cell lines, and biological reagents;

■ Overseeing the networking and distribution of data and biological samples until the regional centers are set up;

■ Assisting in organizing and funding human gene mapping workshops and other international meetings;

■ Assisting international exchange of knowledge and research techniques through training fellowships, instructional courses, and workshops;

■ Offering expert advice to governmental and other agencies on supporting genome research;

■ Producing and distributing a periodic summary of genome activities.

Victor McKusick of Johns Hopkins University in Baltimore, Maryland, was elected HUGO's first president. It was a fitting position for McKusick, who is the "Grand Old Man" of genetic mapping. HUGO's three vice-presidents included Sir Walter Bodmer of the Imperial Cancer Research Fund; Jean Dausset, CEPH in Paris; and Kenichi Matsubara, Osaka University, Japan. John Tooze of EMBO in Heidelberg, West Germany was elected secretary. Walter Gilbert, Harvard, was treasurer. The initial executive committee for HUGO included Charles Cantor (Columbia University and the Genome Center at the University of California, Berkeley); Malcolm Ferguson-Smith (Cambridge University); Leroy Hood (California Institute of Technology); Lennart Phillipson (EMBL, West Germany); and Frank Ruddle (Yale University).

The officers and executive committee members were part of HUGO's 42-member council. Others included Sydney Brenner of the Medical Research Council (MRC) in London; George Cahill, from the Howard Hughes Medical Institute headquarters near Washington D.C.; Renato Dulbecco of the Salk Institute in California; François Jacob at the Institut Pasteur in Paris; Andre Mirzabekov, at the Institute of Molecular Biology of the Soviet Academy of Sciences; Peter Pearson of the Sylvius Laboratories in Holland; Cold Spring Harbor's Jim Watson, and Norton Zinder from Rockefeller University. Five of the forty-two—Dausset, Dulbecco, Gilbert, Jacob, and Watson—were Nobel Laureates. (Only three, by the way, were women: Nancy Jenkins, Mary Lyon, and Elizabeth Robson. In genetics, as it most sciences, female humans continue to be a small minority.) Thirteen nations were represented at the founding council meeting, including Australia, Canada, France, Germany, Greece, Holland, Italy, Japan, the Soviet Union, Sweden, Switzerland, the United Kingdom, and the United States.

THE CEPH COLLABORATION

International collaboration has always been an essential part of all sciences, including biology. Research groups are frequently composed of scientists from different countries, each working in their own laboratories and sharing information and cell cultures. The collaborative nature of biology and genetics continues in the Genome Project. One good example of this is an ongoing international collaboration that began in the Year of Orwell, started by a Nobel Prize-winning Frenchman.

It is called CEPH, an acronym for the Centre d'Etude du Polymorphisme Humain, the Human Polymorphism Study Center at the College de France in Paris. CEPH is headed by genetics researcher Jean Dausset, who won the Nobel Prize for Medicine in 1980. In early 1984, Dausset and CEPH set out to produce a detailed map of genetic markers covering all the chromosomes in the human genome. Dausset is using permanent cultured cell lines from 40 large multigenerational families. The cell lines have been freely donated by researchers from around the world. Many of them came from Ray White at Howard Hughes Medical Institute in Salt Lake City. Most of the families Dausset is using have at least five children, and the majority of them still have all four grandparents still living. In turn, Dausset makes available samples of the DNA from the cell lines to all investigators who are interested, and who agree to abide by CEPH's conditions. Within a year of the beginning of the CEPH collaboration, Dausset had enlisted the collaboration of at least 15 laboratories around the world. The number has since grown to several dozen.

CEPH's conditions are simple, but essential. For example, one of CEPH's conditions is that the laboratory using their DNA must have a DNA probe that can detect the presence of a RFLP. This condition is pretty straightforward. The whole point of the CEPH collaboration is to create a restriction marker map of the human genome. One needs DNA probes that can find and identify new RFLPs to do that.

Another condition is that participating researchers must try to find the inheritance patterns of the new RFLPs in as many of the 40 families as is possible. This means that participating researchers begin by testing the DNA of the parents first. The researchers look for those families in which at least one of the parents has different variants of the RFLP on both members of the chromosome pair in question. These individuals are *heterozygous* for the RFLP. The word comes from the Greek "hetero-", meaning different, and the Latin "zygo-", meaning pair. Those families that do have at least one parent heterozygous for the RFLP will be probed still further, into the DNA of the children and grandchildren. Families in which both parents are homozygous, with no variations in that area of their DNA, are not useful for the project. There are no inherited genetic variations to trace.

Once the inheritance pattern of the RFLP has been traced through all the families, the collaborating researchers must send all their data back to CEPH in Paris. A computer then compares the new data with data from other collaborating laboratories. This speeds up the linking of different RFLPs and the eventual creation of a complete genetic marker map. One notable aspect of the collaboration is that researchers *share only the data*. They do not have to send clones of their DNA probes. This was a smart move by Dausset. DNA probes are like trade secrets. Indeed, in the biotechnology industry they *are* trade secrets. Their creators, even in academic and government laboratories, jealously guard them. By allowing interested researchers to keep control of their DNA probes, Dausset made it possible for them to collaborate safely. He eliminated the possibility that one researcher might steal another's probe. Such theft is not unthinkable. In the high-tech, high-stakes, big-ego world of biotechnology and the Genome Project, the temptation to engage in such genetic pilfering can be nearly irresistible.

Dausset and CEPH also require researchers who are using DNA from the CEPH collection to agree to refrain from using it for commercial purposes. This has the effect of keeping the CEPH collaboration a strictly scientific endeavor. Commercialization of genetic restriction maps may be perfectly legal and ethical. In fact, several corporations are almost solely dedicated to that proposition. However, not everyone approves of such undertakings. Also, not every researcher wishing access to CEPH's collection of DNA may have the financial or political backing to commercialize their findings. This CEPH condition keeps everyone involved on an equal footing.

The CEPH collaboration is also having another side-effect: It is establishing a kind of "standardized" or "reference" version of the human genome. No one person will achieve genetic immortality by having his or her genetic code completely mapped and proclaimed *"the* human genome." Rather, the restriction map of the 23 human chromosome pairs will be created from a whole collection of human genetic sequences. Much of it will be from the 40 anonymous multigeneration families—black, white, red, yellow, American, French, German, Japanese, Mormon, Catholic, Baptist, Muslim, Buddhist, and on and on—whose DNA is permanently preserved by CEPH.

THE VALENCIA CONFERENCE

In October 1988 representatives of several nations met in Valencia, Spain to discuss international cooperation on the Genome Project. The meeting made one thing quite clear. While the United States might be leading the world in time committed and money spent on the Genome Project, other countries do not consider it an "American project." Everyone wants a piece of the Genome Project pie, and feel they deserve it, too. In the end, the gathering produced little in the way of tangible results, but it did clarify the capabilities of different countries for mapping and sequencing the genome. Alexander Bayev summarized the Soviet genome project effort. Yoji Ikawa of the Riken Laboratories in Japan did the same for that country's efforts. The conferees also heard reports on Genome Project efforts in Italy, France, and Germany; a summary of the genome mapping programs being pushed by the European Economic Community; proposals for a UN genome program to be headed by UNESCO (United Nations Educations, Scientific and Cultural Organization); and Victor McKusick's comments on the establishment and goals of HUGO.

All of the international efforts in the Genome Project were obviously dwarfed by those in the United States. It was there that the Genome Project per se had come to life, and the American government had committed more than $50 million in funds for fiscal 1989 alone.

It was obvious that the worldwide Genome Project was going to take off—had already begun in fact if not in name. The major players were clearly going to be the United States and Japan, along with the technologically-developed nations in Europe. Many of the less developed nations wanted to play a role, also, and their representatives made that clear at the Valencia meeting in October 1988. They were also concerned that the First-World nations would have nearly exclusive access to the data to be gleaned from sequencing the human genome. They did not relish a repeat in biology of the exclusion from other advanced technologies that their countries already suffer. They told their colleagues at Valencia that Third World nations and their scientists must participate in the Genome Project in some constructive fashion, and that they must have guaranteed access to genomic data.

In the end, however, the Valencia conference produced little in the way of tangible results. There was no agreement on how to assure Third World nations of unrestricted access to genomic information, or how to coordinate international cooperation on mapping and sequencing the genome. Indeed, in October 1988 the Americans were still struggling to get the Department of Energy and the National Institutes of Health to coordinate *their* genome efforts. Since the United States had not yet gotten its act together, it was perhaps asking a bit much to expect a host of other nations to do it among themselves.

The final, formal declaration of the Valencia conference urged the attending nations to cooperate on the Genome Project. Unlike a preliminary draft, the final declaration did not call for bans on germline genetic research or a gene transfer experiments in human embryos. Valencia was not Asilomar, Norton Zinder of Rockefeller University told the conference. The people at Asilomar in 1975 were a deliberative body picked by NIH. The Valencia Conference was not that at all, and had no authority to call for a ban on anything. The declaration also endorsed HUGO as a preferred international clearinghouse organization, rather than UNESCO. It was all rather vague and undefined. Despite that, many at the Conference thought the declaration would be of some help in future international efforts to complete the Genome Project.

UNESCO AND THE GENOME PROJECT

Despite the lack of a vote of confidence at the Valencia Conference, UNESCO pressed ahead with a proposal of its own. Its goal was to play a central role in coordinating international research. More specifically, the UN agency wanted to focus on the ethical aspects of genome research, and on increasing the role of scientists from Third World countries. In this way, UNESCO perhaps hoped take some of the concerns voiced at Valencia and run with them. UNESCO's Federico Mayor proposed to the agency's 148 member states that it allocate a half million dollars over a two year period to carry out such activities. The money would be used for fellowships and travel grants, so Third World scientists could visit labs in the developed nations doing work on the Genome Project. Some of the funds would also be used to help distribute information about the Genome Project in less-developed nations.

UNESCO got support in its proposal from several members of HUGO, the international Genome Project organization formed in 1988. In early 1989 Mayor established an advisory group for his effort, which included Victor McKusick, president of HUGO; Jean Dausset of CEPH; and Alexander Bayev, who heads the Soviet genome efforts. At the end of June the agency held a meeting in Moscow on the Genome Project, and examined proposals that would be submitted to UNESCO's General Conference in October 1989.

The UN agency's effort to grab a piece of the genome pie was not entirely altrustic, of course. UNESCO had fallen on hard times in recent years, in large part because of the controversial policies of its previous director-general. Western nations became convinced that UNESCO was being politicized, especially by efforts to expel Israel from agency membership and include the Palestinian Liberation Organization in its place. In 1985 the United States withdrew from membership in UNESCO, and Great Britain followed a year later. The withdrawals were a severe blow to UNESCO both politically and economically—especially economically. Not surprisingly, Fed-

erico Mayor—a Spaniard and a scientist—was elected the agency's new head shortly thereafter. UNESCO hoped that by align-
ing itself with the hottest international science project going,
it would regain some of its lost prestige. That, in turn, might
bring England and the United States—and their sizable monetary contributions—back into the UNESCO fold.

THE BRITISH GENOME PROJECT

Great Britain is also playing a significant role in the Genome
Project. This is not surprising, considering England's significant place in the history of medical and genetic research. After
all, it was in the Cavendish Laboratory at Cambridge that Watson and Crick discovered the double helix nature of DNA.

In early 1988 the British government set up a special committee of the MRC in London to coordinate research connected
to sequencing and mapping the human genome. The committee was jointly chaired by MRC secretary Dai Rees and
Walter Bodmer of the ICRF. At the same time, Sydney Brenner
(then at the University of Cambridge) had received funding
for a project to research different ways of sequencing the genome. Meanwhile, an overall strategy for genome research in
the United Kingdom was formulated with the help of several
dozen top British scientists.

By 1989 the British effort was beginning to move into high
gear. It has been spearheaded in large part by Bodmer at the
ICRF and Brenner, who had moved to the Molecular Genetics
Unit at the MRC in London. Brenner, like Paul Berg, felt it
was important to structure the Genome Project in a way that
produced practical results early on. For that reason, the general
scientific strategy of the British effort has been to construct a
genetic map, rather than a physical map, of the genome. The
map Brenner and others envision would concentrate on pinpointing the location of genes that have already identified,
rather than finding new ones. In this way, the British mapping
effort will help researchers who are working on specific ill-

nesses with genetic causes or predilections. The first step is the compilation of information on about ten percent of the human genome. Researchers are working with about a hundred DNA clones, and sending the clones to different groups for further work.

The coordinating center for this and other parts of the British effort will likely be the MRC's Clinical Research Center at Northwick Park in northeast London. It is envisioned as a "genome resource center." Mapping and sequencing projects would take place there. The Research Center would also provide support for genome mapping efforts in other laboratories in the United Kingdom. One major step in this direction was the announcement in 1989 of plans to establish a major computerized database at the Research Center. The database will store and distribute information about the structure and functions of the genome.

In the long run, the Research Center could become one of several nodes or coordinating centers in Europe for the Genome Project. Its database would later be linked with databases at other nodes in Japan and the United States. Brenner believes there will eventually be four or five such centers worldwide, all connected to one another and sharing data freely. To that end, the MRC was spending about $15 million a year in 1989 for genetic mapping. (Other organizations in Great Britain, such as the Imperial Cancer Research Fund, were spending about the same amount.) The MRC's new genome resource center was to be funded with an additional $19 million to be spent over a three-year period on R & D. By 1991 or 1992, Brenner hopes, the Center at the MRC will be in place and fully functioning. Great Britain would then be able to play a major role in international efforts for the Genome Project.

Indeed, the MRC has already made some international moves. It has offered to provide office space and administrative assistance to HUGO. It has also offered a similar arrangement to Japan, which is thinking about setting up a nongovernmental "Human Frontiers Science Program." There could eventually be connections between this and the Japanese genome mapping effort, although that remains to be seen. Some at the MRC have even suggested that the Human Frontiers project might share some administrative and office costs with either HUGO, or the British genome program, or both.

THE SOVIET GENOME PROJECT

Early in 1988, the U.S.S.R. Academy of Sciences announced that it was preparing to launch its own human genome program. In March of that year, Academician Alexander Bayev presented a draft of a human genome project at the Academy's annual meeting. Part of the proposal is to set up an "Institute of Man," which would bring together all the data available on humans as biological and social beings. Bayev urged the Academy to move forward on such a program, arguing that it would have a profound impact on the development of biological science in the Soviet Union.

The U.S.S.R's program is ambitious, and will include genetic and physical mapping projects, sequencing areas of the genome that are of medical interests, and work on sequencing the genomes of other species such as yeast and *Drosophila*. Bayev said the Soviets plan to set up several regional centers for their genome project, but that full-scale sequencing of the human genome would await development of automated DNA sequencing machines. All in all, the Soviet's announced genome project sounded remarkably similar to that of the United States.

Soviet scientists have never been major players in biology or genetics. In a real sense, Soviet biology is still recovering from the notorious theories of Trofim Lysenko. That Russian biologist proposed in the 1930s and 1940s that environmentally acquired characteristics could be inherited. Lysenkoism is just the opposite of Darwinian evolutionary theory, which holds that characteristics acquired in such a manner cannot be passed on to future generations. Josef Stalin decided to support Lysenkoism. It seemed to promise a solution to the chronically disastrous Soviet agricultural harvests, a solution which did not depend on the time-consuming work of improving Soviet crops by standard breeding programs. It didn't work, of course. Lysenkoism is scientifically false. Nevertheless, a whole generation of Soviet biologists bought into Lysenkoism in order to save their professional (and often personal) lives.

Only now, as the Soviet Union begins opening up to the West, are Russian biologists and geneticists beginning to do significant work. For example, Soviet researchers are now publishing solid scientific papers dealing with work on the genomes of some bacteria. They are also working on developing computerized methods for analyzing DNA sequences. The Soviet genome mapping effort may start as something smaller than the "Institute of Man." But it will most assuredly grow, especially if the Soviet leadership redirects government funds from military spending to other areas such as biological science.

FRANCE, ITALY, AND WEST GERMANY

Outside of the United States, Japan, and the Soviet Union the only other nations playing a significant role in the Genome Project are European.

■ In 1988 the French government decided to make an additional eight million francs (about $1.4 million) available for human genome research. Then-Prime Minister Jacques Chirac called such research a new national priority for France. Jean Dausset, founder and head of CEPH, would chair a committee that would allocate the additional funding. Half of it would go to support research into sequencing parts of the genome that have significant medical importance. The rest of the money would go for data processing.

■ The Italian government has made a (for them) major financial commitment to the Genome Project. It plans to spend about $10 million through the early 1990s on a national effort to sequence the X chromosome. Many hereditary disorders are sex-linked. That has made it possible for researchers to identify and locate them on the X chromosome. The result of the Italian project would be a detailed physical map—a "mini-Ultimate Map," if you will—of this important part of the human genome. The project will be carried out by laboratories throughout Italy. Some non-Italian researchers are frankly startled, and skeptical of this initiative. Italian labs are not known for being technologically up-to-date. An indication of the importance the Italian government puts on the project may be seen in who is coordinating the effort: Nobel Laureate Renato Dulbecco of the Salk Institute in San Diego, California, the author of the seminal letter in *Science* magazine in 1986.

He is an Italian by birth.

■ In West Germany, by contrast, not much has happened. From 1986 through 1988, a German governmental agency named Deutche Forschungsgemeinshaft (DFG) had allocated about $500,000 a year for human genome studies. The allocation increased in 1989. However, there were still no centralized research programs in West Germany by the end of the decade.

In addition, much of the European opposition to the Genome Project comes from West Germans, especially those associated with the Green party. The Greens are a coalition of many smaller political organizations focused on environmental issues. They often appear to have an anti-science bias (unfortunate considering the dramatic growth and maturity of ecology and other environmental sciences worldwide). Their skepticism about science has extended to genetic research in general and the Genome Project in particular. Some have suggested that the West German uneasiness about the Genome Project, and the Greens' opposition to it, may be rooted in the continuing German guilt complex about the Nazi atrocities of World War II.

THE EUROPEAN ECONOMIC COMMUNITY

Individual European countries were not the only announced players in the Genome Project. Also in the game was the European Economic Community. The EEC was planning a new program originally called "Predictive Medicine" and later the "Human Genome Analysis" proposal. It would coordinate genome research in different European countries and avoid duplication of efforts both there and with the United States and Japan. The project would have four main goals: establishing a network of clone libraries; increasing the resolution of gene maps; improving and transferring genetic technologies and moving them throughout the twelve EEC member states; and developing integrated database techniques. The funding involved would be small—about $20 million for three years. The program would function mainly as a catalyst for growth and change.

A separate genome mapping project funded by the EEC was also expected to start in January 1989. This six-year, $20 million program was to focus on mapping small non-human genomes. The first was the yeast genome, with about 15 million base pairs. Led by Andre Goffeau of Catholic University in Belgium, researchers in 35 European laboratories would first sequence yeast chromosome 3, with 360 thousand base pairs. Each lab would work on 22,000 base pair segments, and would be paid $5.00 for each base pair sequenced. The goal is to have yeast chromosome 3 completely sequenced by 1991. Meanwhile, American and Japanese researchers are working to sequence yeast chromosome 6, and some Canadian researchers are tackling yeast chromosome 1.

A draft of the Human Genome Analysis program was slated to go to the European Parliament in January 1989. If the parliament approved, the project would start in spring 1989. In January, however, the EEC genome programs hit a serious snag—the West German Greens.

The leader was Benedikt Härlin, a representative of the Greens on the European Parliament. This latter group is a quasi-governmental organization with representatives from different European countries. The Parliament plays an important role in influencing the decisions of the European Commission, the Brussels-based group which in turn coordinates the joint activities of the EEC. Adroit political maneuvering in the Parliament by Härlin and other Genome Project critics threatened to alter the EEC's Genome Analysis program significantly.

Härlin's main objective was to force the EEC to examine and resolve some of the ethical and moral controversies attendant upon the Genome Project and its implications for human gene therapy. Through the Parliament's Energy, Research, and Technology Committee, Härlin introduced a series of amendments to the EEC Genome Analysis proposal. One, for example, called for funding to study "the history of and current trends in eugenics." The word "eugenics" rightly raises the hackles of scientist and layperson alike, with its connotation of forced human breeding proposals and Nazi-like extermination of those "races" with "undesirable characteristics." Another amendment, though, called for the preparation of a list of measures that might prevent the misuse of data about the human genome. A third major amendment would require legal agreements between researchers and those individuals whose DNA was used in mapping and sequencing. Included would be the families whose DNA would be added to the CEPH collection in Paris, in collaboration with the Howard Hughes Medical Institute in the United States. This amendment was clearly connected to the controversial proposal that people whose blood or tissue was used for medical research or experimentation were entitled to some sort of monetary compensation. Some have suggested that these people even receive royalties on drugs or medical treatments that result from scientific research made with their tissues. Härlin's suggestions were thus a mixed bag of the reasonable, the controversial, and the outrageous.

Officials on the commission responded. They noted that they already had a working group set up to examine the moral and ethical issues of the project, and that it was planning at least one public meeting on the subject. Further, they argued that it was impossible in practice for anyone to foresee and resolve every possible ethical dilemma associated with the development of *any* new technological undertaking—including genetic engineering and the Genome Project.

In March the European Parliament voted on the amendments to the program submitted by Härlin and its research committee. Practically all were approved. The commission began revising the Human Genome Analysis proposal. The next twist for the EEC program came at a meeting of the research ministers of the EEC's member nations. Filipo Pandolfi, the man responsible for the European commission's research programs, threatened to stop the amendments to the Human Genome Analysis program. By "stop" he meant "delay." Pandolfi wanted more time to consider further revisions to the program following the passage of Härlin's controversial amendments. That, in turn, meant a delay of at least several months for the program's start.

THE JAPANESE GENOME PROJECT

When American researchers first began seriously considering the Genome Project, one of the goads to make the commitment was the fear that Japan would do it first. There was much speculation in 1986 and 1987 that the Japanese were already committed to a massive genome project of their own. The speculation was fueled by the reluctance of Japanese researchers to talk about their genome projects with others. They would come to international meetings on the subject, ask questions, and take notes. But they almost always refused to answer questions put to them by others, and rarely shared data that had not been already published in scientific journals. Although few scientists would say so publicly, many complained in private about the closed attitude of the Japanese.

At the 1986 NIH meeting on the genome, Japanese researchers had said they were not planning an all-out Big Science genome project. That was not quite true. In point of fact, the Japanese government had had a project running since 1981 with the ultimate goal of developing automated technologies that would sequence up to a million base pairs of DNA a *day* by the early 1990s. Even by the end of the 1980s, that was more DNA than was sequenced worldwide in a year.

By the end of the 1980s, however, Japan had fallen far short of its lofty genome mapping goals. It appeared that the real debate in Japan about the actual value of a genome sequencing project was just beginning in 1989. Nor were there any firm ideas about funding such a project. A report in January 1989 by an advisory group to the government's Ministry of Education, Science and Culture (called Monbusho) endorsed the sequencing idea, but suggested Japan do it in collaboration with other nations. The Science and Technology Agency had issued a similar report a year earlier. Neither agency had offered specific proposals on how to carry out such a project, or how to fund it.

A look at funding levels provides one glimpse at how far *behind* the Japanese genome mapping effort is in relation to the American project. The U.S. government would spend about $50 million on the Genome Project in fiscal year 1989. The Japanese automated sequencing project was budgeted at only about $1.6 million for its fiscal 1989. Even when other related projects were included, Japanese funding for its genome mapping programs fell tens of millions of dollars behind the American funding level.

In fact, by the end of the 1980s, the Japanese project to develop high-speed automatic sequencing technology was itself in trouble. It was begun in 1981 by the Science and Technology Agency, with University of Tokyo biophysicist Akiyoshi Wada picked to run it. Wada felt the project's goal was to make DNA sequencing an assembly-line operation. In essence, a piece of DNA would be put into one end of a system of automated, computer-driven machines. The machines would carry out all the repetitive biochemical tasks involved in base pair sequencing. Out the other end would come a piece of paper printed with the string of bases for the now-decoded gene. Wada's goal was a "super-sequencing" effort of a million base pairs a day. To that end, he enlisted the help of several major Japanese corporations, including Fuji Film, Hitachi Limited, Mitsui Knowledge Industry Co., and Seiko. They were supposed to use their sophisticated manufacturing ability to create the prototype machines for Wada's genetic assembly line.

Things didn't work out the way Wada envisioned them. Seiko developed a new DNA sequencing machine, but decided it wasn't reliable enough to sell abroad. Its machine is used only in Japan, and Seiko's involvement in the Japanese sequencing project is now fairly small. Fuji Film had developed new kinds of gels for genetic sequencing. The gels, however, were too fragile to be shipped overseas, which limited their commercial value. Fuji eventually dropped out of the project. By the end of the decade, nearly $13 million had been spent on the super-sequencing project. The resulting technology could sequence about 10,000 bases a day—about the same capability as automated DNA sequencers developed in the United States during the same time.

One problem with the Japanese genome project may have been its over-reliance on then-current biochemical methodology. Some Japanese researchers have said it might have been better if Wada and his colleagues had first spent some time and money improving those techniques. The next step could then have been the development of new machines that would use the new techniques. In other words, they charged, Wada put the cart before the horse.

Whatever the reasons for the failure to reach the goals of super-sequencing, the goals have been changed. The Japanese effort now aims at sequencing 100,000 DNA bases a day by 1992. Some are happy with that goal and optimistic about the

Japanese genome project. They feel that real progress has been made, particularly in automating DNA sequencing. Others, including Wada himself, are frustrated. They want a massive Big Science project, but can't get it. The companies participating in the super-sequencing project are merely building small sequencing machines for small labs. They refuse to commit to building larger super-sequencing machines until a commercial market develops for them. Wada and others with similar sentiments may be caught in a Catch-22 situation. An international market for super-sequencers may not develop until super-sequencers are themselves developed.

In any case, the Japanese are still involved in the Genome Project. The Japanese government is funding (albeit at a low level) an effort to map and sequence chromosome 21, the smallest human chromosome. The Monbusho ministry is providing modest funding for a drive to sequence the genome of *E. coli*, one of the most valuable organisms in modern genetics. A genome map of *E. coli* is one of the goals of the Genome Project. Riken Laboratories, with support from the Science and Technology Agency, is collaborating with Maynard Olson of Washington University to sequence the yeast genome. That sequence, too, is one of the Genome Project's goals. And Kenichi Matsubara, director of the Institute for Molecular and Cellular Biology at Osaka University and a strong advocate of a sequencing project, is one of the three vice presidents of HUGO.

7

~~~~~~

Collaboration, Controversy, and Computers

*The frontiers are not east or west,
north or south, but wherever a
man fronts a fact.*

HENRY DAVID THOREAU
A Week on the Concord and Merrimac Rivers

As THE Genome Project started moving into high gear, one of the difficulties in mapping gene locations was becoming clear. Researchers began realizing that doing gene linkage studies—the basis for genetic linkage maps showing the relative locations of genes on chromosomes—was more difficult than anticipated. This was particularly true of genes possibly associated with various forms of mental illness or disease. It wasn't even a case of actually finding the genes themselves, but of simply finding their *locations* on a chromosome. At various times, for example, different research groups announced the discovery of a possible gene locus for schizophrenia on chromosome 5; a location for a gene on chromosome 11 associated with manic depression in Old Order Amish people; a gene location for familial Alzheimer's disease on chromosome 21; and a gene connected with major depressive disorders on the X chromosome. In each case, other researchers could not confirm these gene locations in their own studies of other families with the disorders.

There could be several reasons for the discouraging lack of scientific replication of these findings. They include multiple causes for some mental diseases; poor statistical methodology used by the researchers; not enough families in a pedigree, or enough pedigrees; or the misdiagnosis of diseases. In fact, for many, if not most, mental illnesses there is not even firm evidence of any genetic cause. Even if a mental illness seems to run in a family—as schizophrenia and major affective disorders do—there may not *necessarily* be a genetic cause. There could be a gene or genes that may predispose a person to such mental illnesses, rather than cause it directly. Such predisposing genes could be triggered by environmental conditions or emotional/psychological states. Or they may have no genetic causes. It could be something else. For researchers who are looking for such genes, it is like searching for a very tiny needle in a very large haystack, with no assurance that the needle even exists.

The controversy over the genetic linkage of schizophrenia offers a good illustration of the problem facing the Genome Project in particular and genetic research in general. A group of East Coast researchers reported in 1988 that they had found in the families they studied a clear linkage between schizophrenia and chromosome 5. At the same time, a second research team working in Seattle with a different set of families found no such connection at all. Major affective disorders—mood disorders—are another good example. Several good scientific studies point to linkage of major affective disorders to genetic loci on the X chromosome. Other studies, equally rigorous, point either to other locations or to no genetic linkage at all.

Two possible explanations exist for the disparate findings in these cases. First, schizophrenia and major affective disorders could be mental illnesses that have several different genetic causes or predispositions. Secondly, the diagnoses of the family members in the different studies could have been different. The criteria for diagnosing both mood disorders and schizophrenia have changed many times over the last several decades. It would not be all that surprising if research group A were using diagnostic criteria for affective disorders different from those used by research group B. It is thus not surprising that different researchers come up with different results in genetic linkage studies of mental illnesses. Miron Baron of the Columbia University College of Physicians Surgeons has placed the putative gene(s) for major affective disorder near the gene for color blindness, on the long arm of the X chromosome. But other researchers place it on the opposite end of the X chromosome's short arm. And Elliot Gershon of the National Institutes of Mental Health (NIMH) has found no evidence of any X chromosome linkage—on either arm—for major affective disorders.

The same difficulties have cropped up with other reported findings of gene linkages to mental illness. James Gusella of Massachusetts General Hospital gained fame for pinpointing the location of the gene for Huntington's disease, the illness that killed famed folk singer Woody Guthrie. Huntington's is that still-rare case in genetic linkage studies. It is an illness affecting behavior and brain function that has definitely been linked to a single gene, a gene whose location has been found. Its location, at the end of the short arm of chromosome 4, is coded as 4pter-p16.2: between the terminus (end) of the p (short) arm and the 16.2 area on that arm. Huntington's disease is a "gold standard" case for geneticists. The gene is dominant. Researchers have always been able to trace the disease in family pedigrees because its classic symptoms almost always start appearing before the person is forty. Even so, it took decades of effort even to lay the groundwork for the brilliant genetic sleuthing of Gusella and others, who finally found the gene. But he hasn't done quite so well with another gene search. In 1987, Gusella and his colleagues reported finding a linkage between a gene for familial Alzheimer's disease and chromosome 21. Work by other researchers in 1988, however, has not confirmed Gusella's claim.

The ultimate reason for the difficulty in confirming genetic linkages for some mental illnesses may have been noted by Kenneth Kidd at Yale University, one of the world's premier genetics researchers and a leader in the Genome Project. Kidd has pointed out that much of genetic linkage is based on assumptions. Many researchers are *assuming* that there is a gene or genes connected to such-and-such a mental illness. They must. One cannot do a linkage analysis for a gene that does not exist. One cannot create a genetic linkage map for genes that are not there. The resulting map would be about as useful as a map of Atlantis or Oz. In any case, the researchers then try to find a linkage between known gene *markers*—which are certainly present on the chromosomes—and a *hypothesized* gene. If their assumptions are correct, then the linkage process is successful. If their assumptions are wrong . . . .

The solution to this difficulty facing scientists involved in the Genome Project is straightforward: more data. Researchers must find and work with several very large multigenerational families that clearly have a common genetic defect and whose members display similar symptoms. If that is not possible, then they must find and work with many more smaller families with the same criteria. In point of fact, this process has begun. CEPH in Paris is adding DNA clones from new sets of multigenerational families to its previous collection of DNA from 40 families. In the United States, NIMH is searching out large multigenerational families.

NEW FACES AND OLD

As 1988 neared an end, the Genome Project seemed on its way to taking its first steps. The total budget for the various programs was about $50 million for Fiscal Year 1989. James Watson had begun settling in as the new (albeit part-time) associate director for human genome research, heading the Office of Human Genome Research at NIH. He soon received two full-time staff people. Elke Jordan had been the associate director for program activities at the National Institute for General Medical Sciences (NIGMS). Mark Guyer was a geneticist who had also worked for NIGMS. It was perhaps not insignificant that they came from that particular Institute. The NIH's genome office was not an independent Institute of the NIH, and was therefore unable to provide its own grants to researchers. Until that bureaucratic change was made—and there was every indication that it would take place some time in 1989—Genome Project grant money from NIH was coming from the NIGMS.

In another personnel move for the NIH's genome effort, Norton Zinder of Rockefeller University was named to chair the program advisory committee for Watson's genome office.

The eleven other members of the committee included Bruce Alberts (at the University of California), David Botstein, Victor McKusick, Maynard Olson, Phillip Sharp (of MIT), and Nancy Wexler (of Columbia University).

Over at the Department of Energy, Genome Project efforts were also moving into a new phase. Its programs were to receive $18 million in funds during Fiscal 1989. Charles Cantor, the brilliant genetics researcher from Columbia University, was now moving across the country to the Lawrence Berkeley Laboratory at the University of California at Berkeley. There he was taking over the leadership of DOE's Human Genome Center at the lab. DOE also appointed a steering committee for its Genome Project efforts. Cantor was the chair, and the committee included some of the biggest guns in the Genome Project: Leroy Hood of Caltech, Anthony Carrano from DOE's Lawrence Livermore National Lab, Thomas Caskey of the Baylor College of Medicine, and George Bell of the Los Alamos National Lab in New Mexico. Also on the committee as ex officio members were Benjamin Barnhard, the Human Genome Project manager for DOE's Office of Health and Environmental Research, and Gerald Goldstein, the acting director of OHER's Physical and Technological Research Division. Sitting in on committee meetings were several "observers": Diane Hinton of the Howard Hughes Medical Institute (HHMI), Mark Guyer and Elke Jordan from James Watson's Genome Office at NIH, John Wooley of the National Science Foundation, and Sylvia Spengler of the Human Genome Center at the Lawrence Berkeley Laboratory in California.

SETTING GOALS AND PRIORITIES

The DOE's genome steering committee had its first meeting on October 18, 1988. The committee heard that DOE would ask for about $18 million in funding for Fiscal 1989 to support the Genome Project. About $5 to $8 million of that was earmarked for new research initiatives. Close to three dozen proposals were expected to be received by the deadline of December 15.

The steering committee had been envisioned as having an oversight role for DOE's Genome Project work. To carry that out it would use three major methods. Two were long-range, ongoing processes: annual progress reports from individual contractors and research teams; selected presentations by researchers at each quarterly committee meeting. The third was more immediate: a Contractors' Workshop to be held some time in 1989.

The committee decided at its first meeting that it wanted to set up guidelines for all OHER-funded projects on sharing data and biological materials. For example, map and sequence data would go into data repositories and be made available to all researchers at what the committee called "an appropriate time." DNA clones and probes might be deposited as a minimum set and be available from the principal investigator for a specific period of time. Carrano of LLNL was to put together a draft policy for data and sample sharing and present it at the next steering committee meeting.

The business of data sharing led to the matter of computers. Since questions of database formats and the management of computer systems for information sharing transcended agency and department boundaries, the committee urged that some kind of joint task force among DOE, NIH, and HHMI be created to develop specific proposals for three major areas: a common nomenclature for DNA clones and probes; the attributes for a computerized laboratory notebook for genome researchers; and the kinds of characteristics that users would prefer for the final database containing genome map data repository—such as the screen design and categories for questions.

The NIH's advisory committee on the Human Genome met for the first time in January 1989. Watson told the group that the Genome Project is not so much a gargantuan research project as the creation of a resource. In some ways, he said, it is like building a giant particle accelerator for high-energy physics. It will take a long time before it is finished. Like an accelerator, it will continue to produce new discoveries for decades after it is completed. Unlike an accelerator, however, the Genome Project will produce results long *before* it is finished.

Watson sees the Genome Project as a 15-year program. It has already begun, with the steady creation of genetic linkage maps and the development of new technologies. It will proceed, he says, to physical sequencing of more and more of the genome. The genome has already been overlaid with genetic markers—RFLPs and others—with an average spacing of ten centimorgans. Some chromosomes are more densely marked than others. Ray White and others are now pushing hard to complete a five-centimorgan linkage map of the human genome. Watson thinks that a one-centimorgan map will be finished by 1994 at the latest.

The human genome will not be the only one to fall under the aegis of the Genome Project, however. The advisory committee agreed that several other genomes will also need to be included in project-related research and funding. They will include the genomes of *E. coli*, yeast, the nematode, *Drosophila* (the famous fruit fly), and probably the mouse and a plant called *Arabadopsis*. Advisory committee member David Botstein liked this approach. It not only makes organizational sense—gathering together several different genome-mapping programs under one umbrella, so so speak—but it also makes scientific sense. Evolution, Botstein has pointed out, is conservative. As species change and evolve into others over time, the same parts get used over and over. This is just as true on the molecular and genetic level as it is on the more grossly physical level. Indeed, this is what has made possible the great leaps in genetic understanding in the last two decades. For example, researchers working on the genetics of the immune system—people like Philip Leder, Nobel Laureate Susumu Tonegawa, and Genome Project advocate Leroy Hood—performed much of their work on mice, not humans. The immune system genes in mice turn out to be similar in many ways to those in humans. The reason is evolutionary conservatism. Nature uses the same sets of genes, the same "cassettes," as it were, over and over. Thus, by learning about the genes of an animal like the mouse or the nematode, scientists can have a deeper and quicker understanding of similar genes encountered later in humans.

The committee also approved of the idea of focusing on work that would bring a three- to five-fold increase in either knowledge or technology. The Genome Project truly is the "Superconducting Supercollider" of biology, the peaceful Manhattan Project of genetics. If it is to reach Watson's proclaimed goal of essential completion within 15 years, it must encourage research and development programs that will move it as fast and as far as possible, as soon as possible. Thus, programs that promise to increase the speed of physical sequencing will receive much encouragement. So will those that produce databases which can contain the massive amounts of data to be produced and make it easily available to researchers.

One difficulty encountered by the committee during its first meeting related to focus. Should more emphasis be placed on building the "intellectual machine" of the Genome Project, or on producing quick results on "big-name diseases" to please the general public? McKusick favored going after genes for "the biggies," as he called them, like Huntington's disease and cystic fibrosis. Botstein objected that tool-making always got short-changed in biology. It was more important to develop a set of tools for the Genome Project, tools that biologists and geneticists could use for many different tasks. Their job, he suggested, was to build the computer, not write the software.

That approach, however, might well produce nothing but very bad public relations for the Genome Project. Olson wondered what the general public—not to mention the politicians—would think of spending billions of dollars on a project that seemed to be producing nothing but new jobs for postdoctoral biology students. Nancy Wexler finally suggested a compromise. Developing new technologies could be done, at least in part, in combination with work on some genetic diseases. For example, Charles Cantor and Cassandra Smith and their colleagues at Columbia had developed the remarkable new tool called "pulsed field gel electrophoresis," a powerful new technique for use in physical mapping of chromosomes. However, they had never used it on a human chromosome until Wexler and her colleagues gave them samples of chromosome 4. That's the chromosome which contains the gene for Huntington's disease. Similar cooperative research would help both the tool-makers and the disease-hunters.

Another concern of the committee was ethics. The Genome Project may or may not create new ethical questions. It most certainly will cause new and greater attention to be placed on those ethical conundrums already raised by contemporary medical technology and genetic engineering. Watson felt that his Genome Office would have to have money to address those issues. When most people think of genetic technology or the Genome Project, they conjure up Frankenstein's monster and similar visions. They are afraid of genetic knowledge. Watson wants people to see such knowledge instead as an opportunity—a door opening into a future that could be much better than today or yesterday.

Finally, there was the very practical issue of organization. The committee decided that the Genome Office would need to be able to set up research centers and fund their construction. To convince universities to establish a center for genome research without offering them what Watson called "the carrot of new space" was unrealistic. If the Genome Project was to advance, it would need the money to offer those carrots. Such centers would be established at laboratories already doing work related to the Project. One problem would be to find enough labs around the country already doing such high-quality work. In early 1989, there weren't too many. The Genome Project was indeed beginning, but it had a very long way to go.

DIVIDE AND CONQUER?

In April 1989 James Watson did again what he often does so well—started people talking. The most effective way to carry out the massive mapping and sequencing effort of the Genome Project, Watson told a scientific conference in Washington, D.C., might be to assign different human chromosomes to different countries. The French might take on several chromosomes, the Italians one or two more, the English still others. The Soviet Union, which was now talking about spending 40 million rubles on a genome mapping program, could adopt one of the larger human chromosomes. Watson felt that there might be considerable political and economic advantages to dividing the genome up in this way. The appeal to national pride might be useful. For example, he said, the Canadian government might be more likely to sink some money into Canadian researchers' genome efforts if they thought that a particular chromosome was "their" chromosome, a "Canadian" chromosome, as it were.

This division by nation would not keep any scientists in any country from working on whatever chromosome they wished. Instead, thought Watson, it might help foster still more international cooperation. Some specific medical or academic institution in one country might become a clearinghouse for data about that country's chromosome. Researchers in other countries working on mapping that chromosome would send their data to the clearinghouse, which would then store and disseminate it as needed.

Some people liked Watson's idea. International pharmaceutical corporations were becoming conscious of the fact that neither they nor anyone else could carry out a single massive genome mapping project. Used to the fierce competitiveness of the commercial world, they were now beginning to talk cooperation and collaboration—at least in basic research on the genome. Dividing up the effort was something they supported. The next step, said one industry representative, Ralph Christoffersen of the Upjohn Company, might be a big meeting of representatives from all the governmental agencies involved around the world. The purpose of the meeting would be, said Christoffersen, to discuss how to produce the money and political clout needed to divide the human genome by country, and then map and sequence it.

However, not everyone was pleased with the proposal. Great Britain's Sydney Brenner, who had masterminded that country's genome initiative, was sharply critical of his friend Watson's suggestion. Brenner worried that it might presage the "Balkanization" of the genome effort. Instead, he continued to urge the establishment of several regional centers to direct the mapping and sequencing efforts. These laboratories (Brenner envisions three or four to cover the whole world) would act as "hubs," with "spokes" of information going out to other individual laboratories and research groups. One of the hubs, in turn, could act as the central location for overall coordination and direction of the Genome Project.

Watson's idea also received a chilly reception at a meeting of HUGO at Cold Spring Harbor in April. Victor McKusick later remarked that many of HUGO's members (there were 220 by mid-1989) liked the idea of specific research centers acting as hubs for information about specific chromosomes from other labs. However, those HUGO members also felt that the scientific community, not politicians or government bureaucrats, should decide which institutions would be research hubs for which chromosomes.

By the end of May, Watson was claiming that the whole brouhaha over his "chromosome per country" idea was a tempest in a teapot. I was misunderstood, he said. His primary concern was to reduce the costs of the Genome Project for the United States as much as possible. That could be done if other countries made monetary commitments to the Genome Project. He felt that a chromosome was a logical unit of management. If the genome mapping and sequencing effort was to be divided up internationally, why not that way? And besides,

Watson insisted, he never meant to suggest that scientists anywhere couldn't work on whatever chromosome they desired.

The NIH's Genome Initiative Advisory Committee agreed with Watson that the chromosome was indeed a logical unit of management in any scheme to divide the work. However, they weren't so sure about the rest of the proposal. Watson let it slip into limbo.

ENTER THE ROUNDWORM

Watson was much more successful with another genome mapping proposal made in 1989. At the April NIH advisory committee meeting, he suggested that the agency fund a collaborative project between the United States and Great Britain to sequence the entire genome of the roundworm. The committee loved the idea, and unanimously approved it.

The worm's scientific name is *Caenorhabditis elegans*. It is a nematode, a group of worms that have been intensively studied by biologists for many decades. The nematode's biology is thus very well understood. For example, it is known to have exactly 958 cells, and the way each cell divides during the nematode's development has been completely described. For more than two decades Syd Brenner and his research teams at the MRC's Laboratory of Molecular Biology in Cambridge, England have been studying *C. elegans*. Several research groups in the United States have also studying the nematode. Brenner's team had recently begun to map the worm's genome. They began by cutting it into 200 overlapping pieces, the cosmids so beloved of Anthony Carrano at the Lawrence Livermore National Laboratory in California. The entire *C. elegans* genome is about 100 million base pairs long, about the size of a human chromosome. Each cosmid is about 500,000 base pairs in length, roughly the size of the genome of *E. coli*. Thus the *C. elegans* genome sequencing was a logical bridge between the sequencing efforts on the genome of the famous *E. coli* and that of human chromosomes.

Unfortunately, it did not seem that there was enough British funding to carry out the sequencing of the nematode's genome. Thus, Watson's suggestion: Brenner's group, and Maynard Olson and his colleagues at Washington University, would each get $600,000 a year for three years for a feasibility study. Olson, of course, is the leading expert in the creation and use of yeast artificial chromosomes or YACs. The idea is that Olson and his team would use YACs to bridge the gaps between the cosmids. Then the number of cosmids could be reduced to about a hundred, a fairly manageable number for actual sequencing. If the three-year study had a positive outcome, the actual sequencing would begin. It would take about fifty people sequencing a thousand base pairs a day for six years. Watson felt that, if it worked, it would be a fitting conclusion to Brenner's career with *C. elegans*. Norton Zinder, head of NIH's genome advisory committee, thought it would also be an excellent beginning to the sequencing of the human genome. By sequencing the entire genome of *C. elegans*, the "worm people" would go a long way in helping researchers find ways to begin sequencing the human genome.

Just how long it will take to sequence the human genome has been difficult to estimate. Estimates from different researchers range from ten to one hundred years. What is needed for a good estimate of the time to a complete Ultimate Map is some kind of benchmark for DNA sequencing, a reality check, as it were. Two teams of researchers, one American and the other British, have provided such a reality check on what is actually feasible. They presented their work at the Cold Spring Harbor meeting on the human genome in April 1989. Bart Barrell and his team at the MRC in Great Britain have sequenced the entire genome for a virus called *cytomegalovirus*, or CMV. The CMV genome is 230,000 base pairs long. Ellson Chen and his research group at Genentech, in California have sequenced the gene for human growth hormone, which is 70,000 base pairs in length. And both groups did it manually, with no automatic sequencers.

Barrell's group took 12 person-years to sequence the CMV genome. The average rate or throughput was 20 kilobases per person per year, increasing near the end of the project to about a hundred kilobases. Chen's group took 1.7 person-years to sequence the human growth hormone locus. That corresponds to a throughput of roughly 40 kilobases per person per year. Chen has estimated that a skilled technician can sequence about a hundred kilobases per year on the average. However, in real life no one could keep up that pace. Manual DNA sequencing is incredibly boring work. The most skilled technician can actually sequence no more than 50,000 base pairs of DNA a year. Chen had three technicians quit during their project.

Barrell and Chen hold the record for sequencing the largest continuous pieces of DNA, human or otherwise. The volume of their work is useful in the sense that it provides a glimpse at the reality of DNA sequencing speed and accuracy. Sequencing DNA is still an art. Finishing a sequence, for example, can often be difficult. It can take as long to sequence the last one or two percent as it took to do the first 98 or 99 percent. Sequencing human DNA, as Chen's group did, is often trickier than viral or bacterial DNA. It contains introns and many repeated sequences, and regions which are difficult if not impossible to clone. Not having clones to use as DNA probes makes sequencing those stretches extremely difficult.

Using Chen's and Barrell's work as a guidepost, it would take about *30,000 person-years* to sequence the human genome. Then it would have to be done again, several times, to check for accuracy. In order for the Genome Project to be essentially completed within 15 years—the time estimated by most of its supporters—computers and automated machines will have to play a major role. They are already beginning to make their appearance.

THE COMPUTERS BEGIN ARRIVING

Some years ago, Nobel Laureate Paul Berg attended a play called "Breaking the Code." It was about Alan Turing, the brilliant British mathematician who was one of the people who invented the modern computer. During World War II Turing broke the Nazi naval code, which seemed to be a babel of incredible complexity. Turing pulled together a group of mathematicians and they deciphered it. What he did essentially was deduce patterns in the "babel" using mathematical principles.

Berg is convinced that there will be such patterns and information in the genome. Molecular biologists, busy looking at small segments, may well not discern the larger patterns. However, people trained in information theory—the Alan Turings of today—might well deduce such patterns. Berg is convinced that they should have a crack at the Genome Project. He wants to challenge computer scientists. The challenge is compelling: Here is a sequence made of just four symbols, four letters, and it is three billion letters long. Can we find any patterns there that make any sense? Indeed, Berg and others have done just that. The Genome Project is an intellectual arena in which the computer nerds are meeting the biology nerds. They are learning each other's language. They are learning to communicate with one another.

THE CONNECTION MACHINE

In one very important way the computers of today are very much like the first electronic computers of the late 1940s, such as ENIAC. The memory unit and the central processing unit are separate. The data is kept in the memory unit, and must be taken from there and put into the central processing unit to be operated upon. The results are then put back in the memory. This "sequential design" means that the processing operations of the computer must be performed one step at a time. When computers were first being built, the memory and the central processing units were made of different materials, so they had to be separate entities. Today, both memory and central processing units are made of the same material, special silicon wafers. In most computers today, about 90 percent of the silicon is devoted to memory chips. However, the central processing unit is doing most of the work, so the other 90 percent of silicon spends most of the time sitting idle. The most cost-effective solution to this would be to combine processors and memory into one unit. One way to do this is by building a computer fundamentally different from most others: a *parallel processing* computer.

A parallel processing computer uses many small processors, all working at once—in parallel, thus the name. Each processor has its own small memory area. In this way both memory and processing capacity can be used with great efficiency.

There are several different kinds of parallel processing computers available today. They include the Massive Parallel Processor, or MPP, used by NASA at its Goddard Space Flight Center in Maryland, the iPSC series of parallel computers marketed by the Intel Corp, and the HyperCube, developed at the Jet Propulsion Laboratory in Pasadena, California. A parallel processing computer which is attracting the attention of genome mappers is the Connection Machine, or CM, built by the Thinking Machines Corp. in Massachusetts.

The basic unit of the Connection Machine is an integrated circuit containing 16 tiny computer processors, and a device for routing communications from unit to unit. Each of the processors has associated with it 4 kilobytes (4 kb, or 4,096 bytes) of memory space. The computer with which this book was written contains more than a million bytes of memory. In fact, a good pocket calculator has more memory than one of the small processors in the Connection Machine. The *circuit board* contains 32 integrated circuits. There are 128 circuit boards in the Connection Machine, which thus has more than 65,000 of these processors. The entire ensemble of chips and boards is contained in a cube just 1.5 meters on a side.

Each set of 16 processors is connected by a special switch that makes it possible for any two of the processing units in the Connection Machine to be directly connected to each other. This is not done with actual wires running from every processing unit to every other. That would require more than 2 billion wires. Instead, the builders of the Connection Machine have connected the processing units by means of a special pattern of connections. The result is that no processor is more than 12 wires away from any other. The result is a computer with incredible speed and flexibility.

A parallel processing computer like the Connection Machine works on problems in a manner different from that of conventional computers. One way to see this is to compare the way the two different kinds of computers would actually "see" an image. A conventional computer examines an image one spot at a time. The image is represented by an array of numbers, and each number corresponds to the brightness of a particular spot on the image. Suppose a particular image is represented by an array of 256 spots on a side, or 65,536 points overall. A conventional computer must operate on that image one point at a time, since its one central processing unit can carry out only one operation at a time. In order to "see" the image, then, the computer must carry out more than 65,000 separate steps.

A parallel computer like the Connection Machine works differently. Each of its processing units can operate on each spot on the image at once, in parallel. All 65,000-plus points on the image are processed or "looked at" simultaneously. What's more, the 65,000 processors are all in communication with one another. That means that the parallel processing computer can compare the data about each point on the picture in different ways, and do so very quickly. The multitude of processors thus makes the Connection Machine extremely fast—more than 1,000 times as fast in some cases as the most powerful conventional mainframe computers. The many interconnections among the processors make the Connection Machine very flexible.

Genome Project researchers are interested in parallel computers like the Connection Machine. For one thing, parallel computers would be lightning fast at searching huge databases for specific information. Their ability to find a particular pattern—say, a certain DNA sequence—and compare it with all the others in a database will make them invaluable tools for the Genome Project. Parallel computers like the Connection Machine would also be highly useful united to a new generation of automated DNA sequencing machines. They could speed up the sequencing process immensely.

Clearly, advanced computers that process information in new ways will have a significant impact on the Genome Project.

THE HUMAN GENOME MAPPING LIBRARY

However, even conventional computers are helping genome researchers. One example of the way in which computers and computer databases are being used by Genome Project researchers is the Human Genome Mapping Library. The HGML is located at Yale University. Since 1986 it has been maintained with funding from the Howard Hughes Medical Institute. It began more than 15 years ago with funding from the NIH, and thus predates the Genome Project. However, more and more gene mapping researchers are finding the HGML to be an invaluable resource. Before HGML switched from the NIH to HHMI, it had a paid staff of two and a half full-time people. Database searches were usually performed by the staff. Direct access to the Library's databases by outside users was quite difficult.

In 1986, however, momentum began to build toward a full-fledged Genome Project. It became clear to the HGML staff that new genetic data was going to come pouring into the system. It would be necessary to revamp the electronic library to deal with that and with the increased demands for services.

By 1989, HGML had succeeded in its tasks. It is managed by Dr. Kenneth Kidd of Yale. The system now has a menu-driven query system which can be used with ease by subscribers to HGML. The databases now include "pointers" to information contained in the GenBank at Los Alamos. By adding MIM numbers to its entries, HGML has made them much more usable to researchers. Finally, the amount of data stored in the HGML databases more than quadrupled, from about 7,000 records in 1985 to nearly 26,000 at the end of 1988. The number of people directly using the data has risen from a mere handful in 1985 to more than 300 subscribers in over a dozen countries.

The HGML gives researchers online computer access to five different databases. They are called CONTACT, LIT, MAP, PROBE, and RFLP. CONTACT is a list of names, addresses, phone numbers, and electronic contact numbers for nearly 3,000 researchers involved in genetic mapping. LIT is a database of more than 9,000 literature citations in genetics and gene mapping. The information in LIT can be searched by author, keywords, title, journal name, year of publication, gene symbols or MIM numbers, the reference numbers for entries in *Mendelian Inheritance in Man*, and Victor McKusick's ongoing and ubiquitous encyclopedic listing of genes and gene loci. MAP is a database containing information about every genetic locus which has been mapped to the human genome. The MAP database contains one entry for each of the genetic loci which have been mapped on the human genome. MAP entries also include the MIM reference numbers. The database can be searched using MIM numbers, the symbol for the genetic locus, its actual name, its location on the chromosome, and GenBank accession numbers. PROBE contains over 5,600 entries about DNA probes or clones which are used in genetic mapping. The entries in PROBE indicate what kind of DNA probe it is, a symbol for the gene or genetic locus which it occupies, and the name of a researcher to contact for more information. RFLP is a database containing information on restriction fragment length polymorphisms (RFLPs) for more than 1,500 genetic loci on the human genome. The database

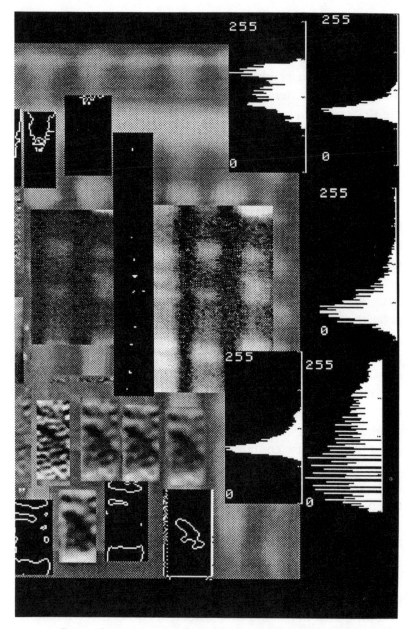

Innovative computer software being developed by researchers at
the Los Alamos National Laboratory will make it easier to compare
different DNA sequences during genome mapping runs.—Printout
courtesy of Tom Marr, Los Alamos National Laboratory.

entries include information on the markers' chromosome location, and the name of the probe used to find the RFLP.

With its current menu system of screens and choices, the HGML is extremely easy to use, even for researchers with little experience in online databases.

BIONET

Another example of a computer network available to Genome Project researchers was Bionet. This was a computer network for biochemists and molecular biologists. It was set up in 1984 by the IntelliGenetics Corporation in Mountain View, in the heart of California's Silicon Valley. Bionet functioned as a nonprofit service to the bioscience community. Its funding came from a grant from the National Institutes of Health, and from individual user fees. Users are charged a subscription of $400 per year for access to the network. In 1987 Bionet had more than 1,700 users representing nearly 500 laboratories around the world. By the end of 1988 the number of users has reached more than 3,000.

Bionet offered its subscribers three major services. First, users got access to a huge volume of information contained in many different databases. Researchers could also make use of a large electronic library of different computer programs. Finally, subscribers could use Bionet to communicate with one another electronically.

Logging onto Bionet gave a researcher access to the latest information in the nucleic acid databases of GenBank and EMBL, the European Molecular Biology Laboratory in West Germany. Bionet users could also enter a databank maintained by the National Biomedical Research Foundation containing information on the molecular sequences of different proteins. Still another protein database available to Bionet users was one maintained at the Brookhaven National Laboratory on Long Island, New York. A researcher at the Cold Spring Harbor Laboratory maintains a database of information on restriction enzymes (the "scissors" used in genetic cutting and splicing) which Bionet subscribers could use.

Besides being able to read information electronically in many different databases, Bionet subscribers could also make use of a library of computer programs. Users were able to "download" the software through the telephone lines from Bionet's DEC 2065 computer to their own desktop PCs. Most of the software available were programs developed by IntelliGenetics for analyzing DNA sequences. They had names like GEL (which could search different DNA computer databases to see if they contained a sequence which matched yours) and CLONER (which could draw a genetic map to use in cloning experiments). Bionet's software library also contained software donated by other biologists and geneticists. In addition, the network also had available a library of free personal computer software commonly referred to as "shareware." The shareware included various computer communications programs and word processing software.

Bionet's third major service was communication. Users could leave messages for one another in one of two ways: either by electronic mail (usually called E-mail) or by means of electronic bulletin boards (or BBSs, for bulletin board systems). Since Bionet users were to be found across the United States, Canada, and Europe, having an E-mail system made communicating much easier and quicker than relying on standard postal systems. Bionet users could also leave E-mail for scientists who were not subscribers, by means of electronic connections to two major computer networks called Arpanet and Bitnet. In fact, Bionet users were able to use E-mail to submit new sequence information directly to GenBank or EMBL databases. Finally, Bionet users could leave a single message for many different scientists by sending it simultaneously to many different electronic bulletin boards.

BIONET GOES DOWN

Not every venture into computerization has been successful. One notable failure was that of Bionet. By the end of 1988 it had received $3.5 million in grant funding from the NIH. According to the original conditions of the grant, the company was to carry out molecular biology research using the computer system as well as provide computer services to people who subscribed to Bionet. However, the agency concluded in early 1989 that the company was not carrying out research as required by the grant's conditions. The NIH therefore decided not to renew the grant. David Kristofferson, Bionet's manager, admitted that the company had not put together a very strong research program, and that the NIH's conclusions about that part of their performance were correct. However, he maintained that this was only because IntelliGenetics was nearly overwhelmed by the demands Bionet's users were putting on the system.

Nevertheless, in September 1989 IntelliGenetics pulled the plug on Bionet, leaving its estimated 3,000 users in the lurch. The hardest hit were researchers at smaller colleges and universities which did not have strong computer facilities. While 3,000 people may not seem like much (some large online networks, like Delphi or CompuServe, have hundreds of thousands of users worldwide), Bionet's demise was a warning to the Genome Project. The Project's organizers have clearly chosen to take an organizational path that lies between massive centralization and traditional cottage industry biology. It is a path that calls for several regional data centers, efficient computer networking, common databases, and simple and effective access to those databases by as many researchers as possible. The computerization of molecular biology and genetics is coming, despite the resistance of some of the old guard. More and more researchers are interested in computers. However, if investigators at small institutions get left out of the computer loop, the Genome Project will suffer. It could also easily become a Science Game for the rich and powerful of the research world, the people who work at at National Laboratories and for giant universities. If the Genome Project is to retain its cottage industry creativity in the midst of Big Biology, failures like the Bionet shutdown cannot become commonplace.

Indeed, systems like Bionet will have to be encouraged and supported. That seemed to be the case in the wake of the Bionet shutdown. At least one researcher, at the Los Alamos National Laboratory, began pulling together support for a follow-on to the system.

THE LiMB DATABASE

One small but useful computer-related tool for Genome Project researchers and others involved in genomic research is a computerized database available through the Los Alamos National Lab in New Mexico. It is called "LiMB," which stands for "Listing of Molecular Biology" databases. The LiMB Database, in other words, is a database of databases. The principle investigator for the project which created LiMB is Christian Burks, one of LANL's computer wizards. John Lawton is the database manager.

In the words of the letter announcing its existence, LiMB "contains information about the contents of [computer] databases related to molecular biology as well as details of how they are maintained." The purpose of LiMB is to make it easy for geneticists and biologists to find and use sets of data that would be relevant to their research.

The information in a LiMB listing is divided into 55 fields, including the name of the database; its location; contact names and phone numbers, including the "addresses" and "names" for electronic mail contacts and the type of information contained in that database.

As an example, consider the listing for the EMBL database, located at the European Molecular Biology Laboratory in West Germany. The listing includes: the entry abbreviation (EMBL); its accession number; the official name (The EMBL Data Library), alternative names (The EMBL Database; EMBL), and old or incorrect names of the database; name, address (postal and E-mail), and phone number for general inquiries, contributing new data, or receiving new data; the source of the data contained in the database; the names of institutions with which the database collaborates in the collection and distribution of data; its source of funding; other databases to which it is cross-referenced; the type of information it contains; the ways in which data can be contributed (on-line; by magnetic tape; on floppy disks; on paper, etc.); the number of entries in the database; how many bytes of information the database contains, and how much it can hold; and a summary of the database's "charter" or reason for existence. In the case of the EMBL Data Library, it is "to collect, organize, document, and make freely available the body of known nucleotide sequence data."

The LiMB Database was first released in November of 1988, and will be periodically updated. Even at this early date, however, it will prove a useful tool to those involved in Genome Project-related work.

SANTA FE 1988

One example of the meetings taking place between biologists and computer experts occurred in Santa Fe in December 1988. The Santa Fe Institute sponsored a five-day symposium dedicated to exploring the interface between computers and genetic sequencing. The meeting brought together nearly a hundred of the men and women working at the cutting edge of computers and genetics. The Los Alamos contingent included Walter Goad, Tom Marr, Christian Burks, Alan Lapedes, and George Bell. Others included Diane Hinton from the Howard Hughes Medical Institute in Maryland; Russell Doolittle from University of California at San Diego; Charles DeLisi from his new position at Mount Sinai School of Medicine in New York; Robert Jones from Thinking Machines Corporation; David Kristofferson of Bionet (which at that point was still operational); Suzanna Lewis and Ed Thiel from the Lawrence

Berkeley Lab in California; and Peter Pearson of the European Commission.

The workshops were highly successful. There was considerable discussion of ways to detect significant patterns in DNA and protein sequences. Thinking Machine's Robert Jones, for example, gave a talk on the use of parallel computers for such searches. There was also discussion of the new super-chip for computers, and work by Lapedes at Los Alamos on using so-called "adaptive networks" to detect protein coding regions in DNA.

Pooling of information was another important theme of the meeting. The three major centers for sequence data about the genome are the GenBank at Los Alamos; the EMBL Data Library in Heidelberg, West Germany; and the DNA Data Bank of Japan, in Mishima. The workshop participants agreed that there needed to be data processing and image processing systems that individual labs can use to organize their own sequence and map information. The labs also need to be able to use such systems to submit new data *directly* to central databases. Everyone also agreed that all genome-related databases need to be connected in some way which will allow everyone access to all of them through individual computer workstations in a lab or office. In fact, the National Library of Medicine and the Center for Human Genome Studies at Los Alamos plan to coordinate just such a computer linkage project.

Thus it appears that the Genome Project is revolutionizing biology and genetics in a way few would have predicted, making possible new methods of scientific collaboration. What biologist of the 1960s—or even the 1970s—would have imagined it?

8

The Forms of
Things
Unknown

. . . .*imagination bodies forth*
the forms of things unknown. . . .

WILLIAM SHAKESPEARE

A Midsummer Night's Dream

F

OR THE GENOME PROJECT, 1989 was a year of
breakthroughs, as the forms of a few "things unknown" began
to reveal their true shapes. It was a year of advances in computers, genetics, and medicine. It was the year that James
Watson began to exercise his considerable talents of organization and persuasion for the Genome Project. And it was the
year that the Project made a small but significant change in
its goals.

THE SUPER-CHIP AND THE GENOME PROJECT

In 1989, Applied Biosystems, Incorporated (ABI), the company
that had licensed Leroy Hood's DNA sequencing machines for
commercial use, announced a new advance in the application
of computer technology to genetic analysis. The company had
obtained an exclusive license to an extraordinary new computer chip. The chip—and ABI's ability to commercialize it—
will soon make it possible for molecular biologists to do work
on their desktop PCs that was once possible only on supercomputers.

In a very real sense, obtaining the actual physical sequence
of the human genome will be the easy part of the Genome
Project. The hard part will be analyzing it: deciphering long
stretches of DNA, determining the patterns that exist, and
translating the patterns into genes. The major problem in such
analysis is that the DNA "alphabet" consists of just four letters
(the nucleotide bases) repeated over and over again. The human genome is "written out" in about three billion letters—
those four DNA letters repeated in various patterns. The quan-
dary is how to obtain the biological information contained in
those three billion letters. The super-chip obtained by ABI will
help enormously in that task. The chip was originally devel-

oped by TRW, Inc., the giant aerospace company, for the Department of Defense.

The connection from TRW to ABI came about through an earlier improbable collaboration, between TRW and Leroy Hood. In 1986 a TRW executive named B.K. Richards happened to attend a lecture at Stanford University on the mathematics of genetics and genetic analysis. There he learned about the great difficulty in extracting biologically important information from the enormously long sequence of DNA that is the genome. Richards realized that TRW's new Fast Data Finder chip might be an answer to the problem. The chip had been designed to scan through the large number of cables and reports that come into the Pentagon each day, and in realtime ferret out the important information in them. The chip, in other words, was more than just an excellent text scanner. It was able to recognize specific patterns of information, and do so with extreme reliability. Richards, a California Institute of Technology graduate, called his alma mater and talked with Tim Hunkapillar of Hood's lab. Hunkapillar came over to TRW's labs to take a look at the superchip and see what it could do. He was almost instantly convinced that it would be a major breakthrough for DNA analysis.

Hunkapillar took one of the chips (which still existed only as a few prototypes) and a Sun 3 minicomputer, designed a DNA analysis system, and wrote some software. The computer program will be available free from Hood's lab to anyone who wants it. The lab in turn put TRW in touch with ABI, which had already collaborated with Hood's lab on the commercial development of his DNA sequencer.

There are several different ways of analyzing DNA sequences for patterns that could be genes. Already so much information is pouring in to different labs that it is impossible to search sequences by hand. Another way is to use current kinds of computers and standard pattern-searching software. However, the kinds of computers most often used today can work only on one problem at a time, in a linear, step-by-step

The DNA sequencer built by Applied Biosystems, Inc.—Photo courtesy of Applied Biosystem, Inc.

fashion. That includes most supercomputers. According to Leroy Hood, even a Cray-2 supercomputer would take more than 5,000 hours (about 30 weeks) to compare all the DNA sequences being uncovered in a year with existing DNA databases. Still another route to take is to develop new mathematical algorithms or formulae faster than the ones in current pattern-recognition software. This is a direction with a great deal of promise, and was widely discussed at the 1988 Santa Fe conference. However, there is still much work to be done.

Hood and Hunkapillar's team has taken still another direction. The group developed a hardware solution to what others have seen as a software problem. They built a machine that is essentially a simple parallel-processing computer. Instead of being limited to working on one problem at a time, going step by step, their new machine can work on many problems at once or on many pieces of one problem simultaneously. The result is that the new DNA analyzer, with its

TRW super-chip, can scan up to ten million DNA characters per second. It is incredibly fast. One researcher at the Los Alamos National Laboratory in New Mexico used the new technology installed in a Sun 3 computer workstation to compare a 10,000 base-pair gene with the 30 million base pairs in GenBank. Its performance was then compared to that of other systems. According to Hood, the comparative analysis took ten days on a VAX minicomputer, one day on a Cray-2 supercomputer—and *ten minutes* with the new super-chip.

The Fast Data Finder chip consists of eight identical microprocessors. Many chips can be put together on computer boards to create systems of practically any size. Different microprocessors on a single chip, or different chips in a system, can be programmed to look for different patterns. The instructions, however, are not contained in software, but rather are hardwired into the chips themselves. This results in an inevitable loss of flexibility, but it also produces a corresponding increase in processing speed. One researcher has compared it to the advantages of using a cake mix instead of a cookbook. Using the mix means you can make only a chocolate cake, and not a blueberry pie. However, you can make the cake much more quickly with the mix than from scratch.

Once the different parts of the system are programmed, the data is fed into it. The raw information flows through the system at a constant rate. The chips do not have to slow the flow down in order to do any number crunching. All they do is look for their preprogrammed pattern. The microprocessors carry out the searching in parallel rather than sequentially. The operator need only "tell" (that is, program) the microprocessors ahead of time which patterns to look for—the sequence that codes for beta-endorphin, for example; or for a certain regulatory region; or to scan a DNA database to see if it includes a base pair pattern like the DNA you have just sequenced.

Unlike most pattern-searching software, the new super-chip is not stymied by complex questions. With software, the more complex the pattern being searched for, the longer the search takes. The Fast Data Finder chip does not need to have complex data preprocessed, as is the case with some programs. Neither does it "walk through" complex searches in numerous steps or iterations. It simply looks for the pattern, no matter how complex it is. Misspellings and vague questions don't slow it down or throw it off. Indeed, the more complex the pattern, the faster the TRW super-chip performs compared to other systems. If one is searching a DNA database for an extremely complex pattern, one need only add additional chips to the system. In essence, it will take the same time to find a pattern equal to a page from an encyclopedia entry about cats as it will to find the word "cat."

Another advantage to the new machines may be cost. Thinking Machines Incorporated's Connection Machine is just as fast as the TRW superchip technology at DNA analysis. However, the Connection Machines cost about $2 million per computer. Some estimates put the cost of the soon-to-come ABI machines at about $40,000 per unit. That's cheap enough for every university in the country to have one.

THE IMPRINTING HYPOTHESIS

As the 1980s drew to a close, biologists and geneticists began to talk cautiously about an idea that would have been considered heretical only a few years earlier: *Not all genes are created equal.* This contradicts one of the major principles of classical genetics, that it makes no difference to a child whether a genetic trait comes from the mother or the father. Since the beginning of the decade, evidence has been accumulating that this might not always be the case. Instead, genes passed on to mammalian children by their fathers might sometimes be different in some way from those passed on by mothers. It now appears that genes from the male parent might be "imprinted" in some way which makes them different from the equivalent genes from the female.

According to Judith Hall of the University of British Columbia in Canada, only about a third of all inherited diseases can be explained by classical Mendelian genetics. The other two-thirds might be understood in terms of the "imprinting" hypothesis.

Each person has 23 pairs of chromosomes, with one chromosome of each pair normally coming from each parent. Standard genetic theory says, however, that what is necessary for a normal complement of genes is two good copies of a chromosome per pair. It doesn't matter which parent each one comes from. So if both copies of a chromosome should come from one's mother, and none from the father, it would not make any difference. But it does. Children who receive two copies of chromosome 7 from their mother can suffer from severe growth retardation, both before and after birth.

Another example is even more dramatic in nature. A child who receives a defective copy of chromosome 15 from its father and a normal chromosome 15 from the mother can inherit Prader-Wili Syndrome, a disease characterized by severe obesity and serious muscle disorders. A child who receives a defective copy of chromosome 15 from its *mother*, however, and a normal one from its father, will not come down with Prader-Wili Syndrome. Instead, the child is likely to suffer from *an entirely different disease* called Angelman Syndrome. It is marked by jerky movements and outbursts of bizarre-sounding laughter. According to Hall, this suggests that the chromosome 15 from the father has an "imprinting" different from that of the mother.

Evidence for "imprinting" also appears in experiments with laboratory mice. Using genetic engineering techniques, it is possible to create a fertilized mouse ovum which has two sets of normal chromosomes from the mother and none from the father. According to classical genetics, this should make no difference whatsoever to the development of the mouse. However, that is not the case at all in real life. Such a fertilized mouse egg suffers from severe retardation of its normal development. It will develop into a mouse fetus if implanted in a female mouse. A healthy placenta does not develop, however, and the embryo dies. It is also possible to create a mouse ovum with two sets of normal paternal chromosomes and no maternal ones. In this case, the placenta develops normally, but the mouse embryo does not. Clearly, says Hall, for mammals to develop normally they must have a set of chromosomes from the mother and one from the father.

Another line of evidence comes from cancer research. In some cases a person is more likely to develop a cancer if he or she loses a chromosome from the mother than from the father. In many cases of Wilms' tumor, a cancer of the kidney, and of a bone cancer called osteosarcoma, it is a chromosome from the mother *rather than from the father*, that has been lost.

The cause of this "imprinting" (if it in fact exists) is not yet known. However, it may be connected to the action of the mother's immune system. A developing fetus grows for nine months inside a woman's body. It changes from being a part of the woman (an unfertilized ovum) to something that is eventually a biologically separate organism—a growing human fetus with its own slowly-developing immune system. The mother's immune system would ordinarily attack and destroy any object or living tissue within the mother that is "not-Self." That's the job of the immune system. In the case of a developing fetus, however, it must not. The fetus must somehow be made invisible to the mother's immune system. This requirement is unique to mammals, by the way. Other classes of animals like birds and amphibians do not grow the fetus inside the mother. The process of protecting the developing fetus from the mother's immune system might somehow imprint or chemically modify the fetus' paternal chromosomes— which come from the father's sperm, tissue alien to the mother's body.

NEW WAYS TO MAP

At the Cold Spring Harbor conference on the human genome in May 1989, two groups of researchers announced they had come up with two new ways of mapping chromosomes. The breakthroughs promise to speed up the creation of such maps. Neither, in truth, is "new" in the sense of "never seen before." One is an improvement on a standard mapping method. The other is the rediscovery of an idea first proposed in the 1970s but then forgotten. The first was developed by David Ward, Peter Lichter, and their colleagues at Yale University, the second by David Cox and Richard Myers of the University of California at San Francisco. The two techniques are very different. Ward and Lichter use fluorescent dyes attached to DNA fragments. Cox and Myers tear chromosomes apart with x-rays, and then make maps of the breakpoints.

The Ward/Lichter mapping method is a new version of the standard in situ hybridization technique. The typical way of doing in situ hybridization is to take a DNA probe and label it with a radioactive molecule. Then the probe is dropped into the chromosome sample to be mapped and allowed to connect up with (to " hybridize" with) it. The probe will seek out and bind to its complementary DNA sequence in the chromosome. The radioactive tag on the probe allows researchers to "see" it, and thus determine the location of its complementary DNA sequence on the chromosome. This is a tedious process that involves placing photographic paper or a plate on the sample and allowing the sample to take a picture of itself. This can often take days of exposure. Laboratory workers must also take certain common-sense precautionary measures since radioactive isotopes are involved, even though the molecules have very low levels of radioactivity and short half-lives.

What Ward and Lichter have done is replace the radioactive tag with a nonradioactive molecular dye. This tag can be detected with fluorescence. Basically, it glows in the dark like a firefly instead of like a radium-dial watch. Gone are the complexities of working with radioactive isotopes. Researchers can actually use a standard light microscope and see the DNA probes as bright yellow-green dots against the red-stained chromosome. Gone, too, is the long delay while photographic plates are exposed to the sample. The new technique's big advantage, Ward has said, is speed. The process of mapping a gene will no longer take months, but will be done literally overnight. During one six-month period that ran through April 1989, three people in Ward's laboratory determined the order of about one hundred probes and genes on chromosome 11.

The Ward/Lichter mapping method is also more accurate than the older in situ hybridization technique. The standard method could locate a probe or gene to within ten million base pairs, a large fraction of the actual length of some chromosomes. The new technique is at least ten times as accurate. With further improvements in equipment, Ward has said he expects to be able to resolve DNA probes only 30,000 base pairs apart on a chromosome.

And there are further improvements in the wings. Early tests of the new technique used only one DNA probe at a time. However, as many as eight probes can be used at once, each marked with a different fluorescent molecular tag and emitting a different color. Ward believes that with the right equipment and about a half dozen people, his lab at Yale will eventually be able to map 4,000 to 5,000 genes a year. At that rate, Ward's lab alone could create a map of the genome with at least one centimorgan resolution in about 20 years. Four such teams working together and divvying up the chromosomes could do it in five.

The second new mapping method is actually based on an idea proposed more then a decade ago by researchers Henry Harris and Steve Goss which was promptly ignored by nearly everyone in the field. David Cox and Richard Myers of the University of California at San Francisco rediscovered the idea and found a way to make it work. They call it *radiation hybrid mapping*. It essentially uses techniques from both genetic linkage and physical mapping methods. In genetic linkage mapping, researchers basically determine how often two genetic markers (the RFLPs) become separated during meiosis. The Cox/Myers method does not rely on that cellular process to separate markers. Instead, they separate the markers by bombarding a chromosome sample with X rays and breaking it apart. The resolution of the map created by this process depends on how frequently the chromosome is broken. That in turn depends on the X ray dosage. Conventional genetic linkage mapping will soon produce maps with about 1 million base pair resolution. Cox and Myers believe that zapping a chromosome with one thousand rads of x-radiation will create breakpoints every 50,000 thousand base pairs. That's an enormous increase in resolution, comparable to that of cosmid maps created with physical mapping methods.

Maynard Olson, the "Yeast Man" at Washington University in St. Louis, is pleased with the development of these two new mapping methods. He believes that they will quickly make it possible to produce a complete low-resolution map of the entire human genome. By "quickly" he means by 1993 or 1994, or even earlier. That will then make it possible to move ahead with the completion of an even higher-resolution cosmid map much sooner than many had thought possible. David Ward agrees with this assessment. His long-term goal is what he has called "saturation hybridization." He wants to map the location of 5,000 to 7,000 DNA probes across the entire genome, spaced about a million base pairs apart, to create a genome map with less than one centimorgan resolution. Ward would then make the DNA clones available for use in creating a high-resolution cosmid map of the genome. The Cox/Myers method would also create a continuous genome map with degrees of resolution that depended on the levels of X ray dosage used.

A GENE-MAPPING MARATHON

Researchers used current methods to track hundreds of genes to locations on specific chromosomes in 1989. In one day, in fact, more than 700 genes were mapped to places on chromosomes. The "gene-mapping marathon" took place on Saturday, June 17, 1989, the last day of the Tenth International Workshop on Human Gene Mapping at Yale University. These massive workshops have been taking place since 1973. They have been characterized as the periodic chromosome housekeeping sessions of the world's geneticists. The researchers essentially try to "tidy up" the existing genome maps, determining the correct places on the 23 pairs of human chromosomes for all the newly-discovered genes. The last such workshop had taken place only two years earlier. The task faced by the 650 biologists at the Yale meeting was enormous. The rate of genetic discovery is now doubling every two to three years. The number of genes with known sites on the chromosomes was now at 1,700. Even so, that comprises less then two percent of the estimated 100,000 genes in the human genome.

In order to carry out the arduous task, the workshop participants broke up into smaller committees, each of which was assigned one chromosome. Their job was to evaluate the reports on newly discovered genes which were tentatively assigned to their particular chromosome. They often had to decide which of two or more conflicting reports made the best case for a gene's chromosomal location. The Yale meeting demonstrated vividly the extent to which computers are now playing a role in genome mapping. Each of the participants was given access to a personal computer linked to Yale's mainframe. The researchers used their computers to talk with one another, input data supplied on floppy disks by the scientists attending, and exchange that data with others online.

At the final session of the meeting, on June 17, each committee presented a report summarizing their findings and location selections. Certain chromosomes stood out. The X chromosome, for example, had to have two different committees working on it, one for each arm. It ended up having more than 100 new genes assigned to it. Chromosome 11 had 34 new genes assigned to it; chromosome 12 got 26. Chromosome 7, which is now known to carry the cystic fibrosis gene, had 71 new genetic markers mapped onto it. Chromosome 4 is known to be the location of the gene for Huntington's disease. That particular gene locus alone had 36 new genetic markers laid in. There were also several chromosomes which were barely touched. The Y chromosome was one, as was chromosome 13. The committee looking at chromosome 18 recommended it as a chromosome on which researchers could work without much worry of competition! The gene-mapping marathon produced some interesting surprises, too. A gene that codes for red hair, for example, was mapped to chromosome 4. The gene locus for Prader-Wili syndrome was mapped to chromosome 15, and that of polycystic kidney disease to chromosome 16. A big surprise was the discovery of a gene for an ailment called Gerstmann-Strassler disease on chromosome 20. The disease was long thought to be caused by a virus or bacteria. Instead, it is known now to be a genetic disease. A much more precise location for the gene causing Down's syndrome was identified on chromosome 21, which was also assigned a locus for a possible gene for familial Alzheimer's disease.

In all, some 53 genes known to be associated with some kind of human disease were mapped during the Workshop. They included

■ a gene on chromosome 3 which causes a form of kidney cancer, and which is also involved in small-cell lung cancer;

■ a gene on chromosome 6 which may play a role in disorders of the body's connective tissue, including rheumatoid arthritis;

■ another gene on chromosome 6 which accounts for fifty percent of the genetic susceptibility for insulin-dependent diabetes, a form of diabetes strongly suspected to be caused by a malfunctioning immune system;

■ a gene on chromosome 11 that codes for a nerve cell receptor for the chemical dopamine; it may be very important in the development of treatments for depression.

THE GENOME PROJECT EXPANDS

The NIH's advisory committee on the Genome Project had its second meeting at the end of June 1989, in Bethesda, Maryland. Besides approving Watson's proposal for a joint American/British project to sequence the nematode genome, they also tried to pull together a "National Plan" for genome mapping research. It focused mainly on the overall goals of the Genome Project, on ethical issues, data handling, and its scientific and administrative organization. The plan would be submitted to Congress for its approval. James Watson wanted to establish a clear U.S. policy on the Genome Project, and have that policy in place by the spring of 1990. The plan's ultimate purpose was to argue effectively for NIH's proposal to spend $200 million a year for fifteen years on mapping the human genome.

Even before the National Plan was submitted, though, it was clear that there was plenty of support for the Genome Project. The seemingly impossible began taking place. Instead of getting its budget requests cut, the Genome Project was being increased. The Department of Health and Human Services had first increased NIH's original budget request for Fiscal Year 1990 to about $62 million. Then the NIH budget disappeared into the forbidding maw of the Office of Management and Budget. For years the OMB has been the terror of federal science funding requests. It has slashed proposed budgets by large percentages in some cases and attempted to eliminate others entirely. With the NIH's genome initiative request, however, the miraculous happened. It had come out of OMB larger than it went in—boosted to $100 million for 1990.

That money might well be put to even better use than anyone expected. At a scientific conference in Philadelphia honoring Sweden's Nobel Foundation, James Watson noted that the costs for sequencing the genetic code were likely to drop. He speculated that it might be possible to bring the costs down to 50 cents per base pair with no major technological breakthroughs. At the same conference, Smith, Kline &

French's George Poste suggested that the completion of the genome map would lead to the production of "a protein dictionary." Such a "dictionary" would be a listing of the proteins produced by each type of cell in the human body. A protein dictionary, Poste predicted, would be a valuable tool in the battle against different diseases.

Meanwhile, other U.S. government agencies were getting into the act. The Department of Agriculture (USDA) and the National Science Foundation (NSF) both announced in mid-1989 their intentions of playing at least a modest role in the Genome Project. Their common theme was botany. Plants.

The USDA's interest in plant genetics is quite logical and has a long history. The agency, for example, supports seed banks and plant tissue banks in different parts of the country. The USDA already has an Office for Plant Genome Mapping Research. Its 1989 budget was only about $100,000. However, Jerome Mitsche, the Office's director, told the NIH's Genome Initiative advisory committee that the agency was interested in boosting the Office's budget a thousand-fold, to $500 million over five years. Exactly what the money would be spent to do, however, was somewhat fuzzy.

The National Science Foundation, on the other hand, has been more specific about its plans for plant genome mapping. DeLill Nasser, the director of NSF's Program for Genetic Biology, organized a workshop at James Watson's Cold Spring Harbor Labs on July 20, 1989. The workshop's purpose was to go over plans by the NSF to map and sequence the genome of a weed called *Arabidopsis thalania*. *Arabidopsis* is a very small plant. It is easy to grow large numbers of it on a single laboratory petrie dish. It also has a relatively small genome, about 70 million base pairs, with very little repetitive DNA. That makes it a good candidate for genome mapping and sequencing. Indeed, it has already begun. Two genetic maps had already been created by 1989, and a physical map was in the process of being produced.

The new NSF project would complete the mapping of *Arabidopsis* for a cost of about $35 million. The mapping project, in turn, is just one of four major goals of the NSF's genome-mapping efforts. Those four goals are

- a central repository for seeds;
- a central repository for specific plant clones and plant DNA;
- a center devoted to completing the physical maps of *Arabidopsis*;
- a center which would sequence cDNA from *Arabidopsis* and ultimately sequence the plant's entire genome, creating an "ultimate genetic map" of *Arabidopsis thalania*.

THE CYSTIC FIBROSIS GENE

By the end of 1989 the Genome Project was beginning to bear fruit, bearing out Paul Berg's vision. Ray White and his colleagues were creating coarse but useful RFLP genetic maps of different human chromosomes, publishing their data one chromosome at a time. Anthony Carrano had begun automating additional steps of the process of producing cosmids and contigs, pointing the way to increased use of robots and automation in the creation of cosmid maps of different chromosomes. Collaborative Research was continuing work on their RFLP map of the entire genome, hoping to fill in some of the large holes. They were also persisting in their efforts to create new genetic diagnostic tests for commercial sale. Researchers were using genetic mapping techniques to discover and isolate the genes responsible for Huntington's disease, retinitis pigmentosa, multiple sclerosis and cystic fibrosis. Efforts continued to find the genes connected to many other diseases with genetic causes or tendencies. And in one of the great stories of modern scientific discovery, a team of researchers using the techniques and tools of the Genome Project finally tracked down the gene responsible for a disease that kills thousands of children and young adults.

Cystic fibrosis, or CF, is the most common fatal inherited disease in North America. The gene for CF is recessive, and 12 million Americans carry one copy of it. If two carriers pass on their defective genes to a child, that child will come down with cystic fibrosis. About 30,000 people in the U.S. have CF, and 1,200 more cases are diagnosed each year. The disease is characterized by a thick mucus which clogs the lungs and makes breathing very difficult. The mucus is also a haven for bacteria, and lung infections are common. The disease is usually diagnosed in infancy because of the repeated infections. The sweat of CF patients has a distinctive odor caused by chemical changes. Finally, CF leads to failure of the pancreas, a gland that produces secretions important to food digestion. Malnutrition is thus a common side-effect of cystic fibrosis. The therapy for CF includes physical therapy to loosen the viscous mucus in the lungs, antibiotics to fight the lung infections, and different enzymes to counter malnutrition. Some CF patients now live into their 20s and 30s, but the disease is eventually fatal.

The search for the CF gene had gone on for years, but by 1984 the gene-mapping work of Ray White and Helen Donis-Keller began to make a substantive difference. The pace began to pick up as research teams around the world worked feverishly to track down the location of the CF gene, and determine the protein it makes. They included Robert Williamson at St. Mary's Hospital in London, Lap-Chee Tsui, Manuel Buchwald and later Jack Riordan at the Hospital for Sick Children in Toronto, Francis Collins of the Howard Hughes Medical Institute at the University of Michigan, Ray White of HHMI at the University of Utah, Helen Donis-Keller's team at Collaborative Research in Bedford, Massachusetts, and others. The story of its discovery was one of politics, personalities, dead ends, fresh starts, hard work, and final triumph. It would make a great movie.

White and Donis-Keller had not been searching for the CF gene themselves at first. But their increasingly large collections of RFLP markers became more and more useful in others' search for it. White was supplying DNA probes to anyone who asked. CRI was not, treating their probes as proprietary information. Tsui and his mentor, Manuel Buchwald, were using probes from White to try and find the CF gene, but having little success.

At the end of 1984, Collaborative Research struck an agreement with Tsui and Buchwald in Toronto to combine the company's DNA probes with the researchers' family data in a search for the CF gene. Tsui set to work, and by August 1985 had found a genetic linkage between one of CRI's probes and the CF gene. They still did not know which chromosome the DNA probe was connecting to. However, CRI told Tsui to hold off, since the company had an agreement with French researcher Jean Frézal to map the location of some of the company's probes, including (presumably) the one being used by Tsui. He was to wait.

Nearly a month later there were indications that the gene was on chromosome 7. However, CRI would not confirm that, despite Tsui's requests for information. Frustrated at the delay, Tsui mapped the location of the probe himself. It was indeed on chromosome 7. CRI was not happy with his actions. When the two groups began working on the research paper to be published in *Science*, the company insisted on reporting that the gene was linked to "an unmapped probe." They won the argument. By October 1985 the science grapevine had already spread the word that the gene was on chromosome 7. However, CRI still refused to say so *publicly*, even at a major genetics conference that month.

Meanwhile, Ray White was beginning to feel upset with Collaborative Research. He had been involved in talks with the company about a joint research effort to find the CF gene. He was surprised and disconcerted to hear of the genetic linkage found by the Tsui-Collaborative collaboration. White later said he felt "misused" by CRI.

Tsui thought the company wanted to keep the location hidden from others until they had more markers mapped onto chromosome 7. At this point the gene could be pinpointed to within only 15 million base pairs. That was not enough to create an effective diagnostic test. Helen Donis-Keller, for her part, has always insisted that the chromosomal location of the CF gene was not fully confirmed until later than Tsui maintained. The truth is that the company was following standard commercial procedures in its efforts to find the CF gene. It had invested millions of dollars in developing genetic probes, and rightfully felt that it at least deserved a fair chance to make a profit from them. Giving away trade secrets to competitors is not a good way to make a profit.

CRI also needed to learn a little diplomacy. At one point Orrie Friedman, the company's chief executive officer, was quoted as saying "We own chromosome 7." By some accounts, they have improved their diplomatic and public relations abilities in recent years.

In October 1985 the Toronto-Collaborative group was ahead in the race to find the CF gene. Within two months their major competitors had caught up with them: Williamson in London and White in Utah. When White first heard the rumors that the CF gene might be on chromosome 7, he and his team immediately began testing every chromosome 7 probe they had. They soon found a very tight linkage of one of their probes to the CF gene—within a million base pairs, 15 times closer than the CRI probe. White has never made any secret of the fact that he made his move because of the rumors. He sees nothing wrong or unusual about that. Williamson has claimed that the rumors played no part in his work. Instead, his team's research had been eliminating different chromosomes from consideration. After a meeting in Helsinki about the cystic fibrosis search, they believed there were only three candidates: chromosomes 7, 8, and 18. He soon eliminated 18 from consideration. Papers by all three groups were simultaneously published in the same issue of *Nature*.

The next episode in the search for the CF gene took place in April 1987. It was an outgrowth of a conference in Toronto convened by the Cystic Fibrosis Foundation. Seven of the major research teams racing to find the gene agreed to pool all their data for collaborative study. The result was a paper authored by Jean-Marc Lalouel of Ray White's lab, and published in December 1986. The cumulative evidence appeared to show that two genetic markers were tightly linked to the still-undiscovered CF gene on chromosome 7. One marker was called J3.11; the other was an oncogene (a gene that causes a cancerous tumor) called *met*. They seemed to lie on either side of the gene. The region encompassed was about a million base pairs in length. Used together, the two DNA probes for these markers could be used for prenatal diagnosis of CF with an 99 percent accuracy. Collaborative Research immediately began working on a commercial diagnostic test for CF based on the two markers and made it available in fall 1986.

White and Williamson, meanwhile, continued the search for the precise location of the CF gene. Because the two markers were a million base pairs apart, they were too far to "walk" toward the gene. One way of *chromosome walking* involves taking small fragments of DNA sequences, beginning at each marker. The fragments are laid out so they overlap at their ends. Researchers then start comparing the sequences, "walking" down the fragments from each marker toward the center. Eventually the gene is found and isolated. However, "walking the sequence" works only for stretches of DNA less than about 200,000 base pairs long. Beyond that, the number of overlapping DNA fragments needed becomes so huge that the amount of time required to sequence and then "walk" is prohibitive. In the case of the two possible CF markers, there was another problem as well. No one knew which marker was on which side of the CF gene. Even if they wanted to "walk the sequence," no one really knew in which direction to begin "walking."

The problem seemed insoluble, but there was a risky short-cut. Human chromosomes with activated copies of the *met* oncogene could be gene-spliced into mouse cells. The mouse cells that later became cancerous would obviously contain the human *met* oncogene. These cells could then be tested to see if they also included the J3.11 DNA marker, and those that did would obviously contain the CF gene. The process by which the *met* marker was activated would reveal on which side of the CF gene it lay—and which way to "walk." The procedure was risky, however. There was some evidence that the process that switched on the *met* oncogene would so mix up the region around it as to render it useless as a marker. It might even get completely separated from the CF gene during the splicing process from human to mouse.

Robert Williamson decided to try the *met* approach. He had already begun the experiment when the problem with the possible flaw became known. He pressed on, hoping to luck out.

He did not. Williamson and his colleagues published their results in the April 7, 1987 issue of *Nature*. They announced the discovery of a "candidate gene" for CF. Williamson was sure he had found it. So convincing was his public posture of confidence that other research teams stopped their search for the CF gene. One was Ray White and his colleagues. As time went by, however, the vagueness of some of their results was not cleared up. Many scientists began getting concerned, then irritated, and finally downright suspicious. White was pushing his friend Williamson for more details, and they were not forthcoming. Helen Donis-Keller—who was not a friend of Williamson by a long shot—was angrily insisting that he release his sequence data.

The situation for Williamson was particularly painful because, by the time the *Nature* paper was published, he had already started finding evidence that it was wrong. He had sequenced two copies of the candidate gene for CF, one from a normal cell and one from a cystic fibrosis cell. If it were the correct gene, there should have been a difference between the two. There was none. By September 1987, at the giant Human Gene Mapping workshop in Paris, Williamson was close to admitting defeat. A month later, he finally did so. The "candidate gene" Williamson had sequenced, now called IRP, was a marker that was probably very close to the actual gene. It might be within 40,000 base pairs.

It was an embarrassing episode for Williamson, and frustrating for people like White, Donis-Keller, and Lap-Chee Tsui. "Working on the CF gene taught us to be careful that the scientific opportunity really is there," White would later say. "We could have been in same position as Williamson."

"We lost about six to eight months of work on it. It takes a lot of effort to gear up again for a search. We finally made a decision not to make the CF search a major focus of our work, and to go on to other things." Other researchers, including Tsui, Williamson, and the Collaborative Research team, continued the search.

Lap-Chee Tsui had begun a collaboration with Jack Riordan at the Hospital for Sick Children. At HHMI at the University of Michigan, Francis Collins was also searching for the CF gene. In 1987 they all knew they had to examine about a million base pairs of DNA on chromosome 7. Collins used a technique called *chromosome jumping*. He took segments of the chromosome about 100,000 base pairs long, comparing them with DNA taken from people with CF, and looking for the closest matches possible. Tsui used a method called *saturation mapping*. He took 250 random pieces of DNA from chromosome 7, compared them to DNA from CF carriers, and consistently found two segments that seemed to match.

At that point the three scientists combined forces. By January 1989 they had narrowed their search to a region of about 300,000 base pairs in chromosome 7. It was the same region that Williamson and his team were examining. Like William-

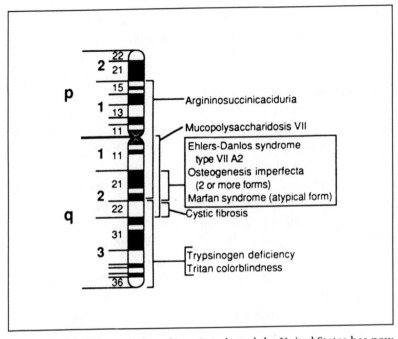

A team of researchers from Canada and the United States has now found the location of the gene which causes most cases of cystic fibrosis. It is on the q arm of chromosome 7.—Illustration reproduced from "Morbid Anatomy of the Human Genome," Victor A. McKusick (1988, Howard Hughes Medical Institute.)

son, Tsui, Riordan, and Collins felt they were also very close to it. Where once they had jumped along chromosome 7, now they began chromosome walking. They used several techniques. One was to cut the DNA into small fragments, tag the fragments with a radioactive isotope or a fluorescent dye, and then see if the DNA fragment made RNA. A DNA segment that makes RNA, which in turn makes amino acids and a protein, is by definition a gene. Of course, they didn't know exactly where to cut the DNA, so the process was random. Also, up to 85 percent of human DNA is "inactive" or "junk," so most of the fragments were nonfunctioning. The process was a bit like looking for a buried treasure (the CF gene) located on one of many tiny islands (pieces of DNA) in a river (chromosome 7). Tsui, Riordan, and Collins were crossing the river by hopping from one island to the next, not knowing which held the treasure.

At the end of August 1989 they found the right island, a segment of DNA that seemed almost certain to contain the CF gene. Now they wanted to determine what protein it made, for that protein is what is immediately responsible for the effects of cystic fibrosis. One of those effects involves chemical changes to the sweat glands. The three researchers created a special library of RNA found in sweat gland cells, and compared their canditate gene with the entries.

And they hit the jackpot—an exact match between their DNA segment and a gene in the RNA sweat gland library which made a membrane protein in sweat glands. A particular mutation of the gene is responsible for 70 percent of all CF cases. The other 30 percent of cases are caused by several other mutations of the same gene. They had found the treasure, winning the race to find and identify the gene for cystic fibrosis, and describing the protein that is the immediate cause of CF. This double success is one of the major achievements of 20th century genetics, and a major accomplishment for techniques directly linked to the Genome Project.

Clearly, Paul Berg's desire to see swift practical results from research related to the Genome Project is beginning to be fulfilled. One result of this discovery will be a treatment for people with CF. Only one copy of the normal gene is necessary for healthy lungs and normal mucus, so it will be necessary to replace only one of the two defective genes in CF patients with a normal copy. One way to do that would be to use a genetically-engineered virus. A copy of a normal human CF gene would be spliced into a special virus that would "infect" lung tissue. Patients could inhale the virus using an aerosol. Many of the viruses would eventually find their way to chromosome 7 and insert themselves into the chromosome. There the normal human gene they carry would replace one of the defective genes in a sufficient number of lung cells, begin producing a normal form of their specific protein, and thus reverse the effects of cystic fibrosis. The carrier virus would die after delivering its genetic package. The Cystic Fibrosis Foundation is already funding research for just such a genetic therapy for CF.

REVISING THE GOALS

The Department of Energy and the National Institutes of Health began developing a national plan for mapping the human genome. The April 1989 meeting of DOE's Human Genome Steering Committee included Watson and Elke Jordan of NIH as active observers. (The October 1988 memo of understanding between the two agencies calls for such joint participation). At the meeting the DOE and NIH agreed to assemble a joint planning committee to work on the national plan. The committee met again in August and in October 1989 to write, revise, and approve the plan for submission to Congress in February 1990. At the same time, the two agencies planned to merge their two groups working on computer and database issues of the Genome Project. Such a move would encourage still more cooperation on an area vital to the Project's success.

By the end of the year, however, the Genome Project itself had hit a snag.

Almost from the beginning, gene mapping researchers said that the short-term goal was the completion within five years of a genetic linkage map with one-centimorgan resolution. Every major report on the Genome Project repeated that goal, including the five-year plan that was being drafted for Congress by the NIH and the Department of Energy. It was predicted that such a genetic map could be finished within five years at a cost of about $15 million per year.

The cat was let out of the bag in December 1989 at a meeting of NIH's genome advisory committee. Washington University's Maynard Olson openly questioned the feasibility of reaching the one-centimorgan map goal. He was not doubting people's desire for such a map—but he asked the committee if anyone really thought it could be done within five years. David Botstein and Lee Hood agreed with Olson. NIH's Center for Genome Research had not been aggressively pursuing the map, they asserted. What's more, the researchers actually doing the mapping were being distracted from the one-centimorgan map goal by the lure of finding specific genes for specific diseases. Most of NIH's grant money was going to efforts to map regions already known to contain disease genes. A more global strategy of blanketing all the chromosomes with markers—the best way to create a genetic map with the desired resolution—was being financially ignored.

Elke Jordan, James Watson's deputy director for NIH's genome center, felt that the charges were a bit exaggerated. She claimed that many researchers were working on the genetic map, although not at the pace or scale originally suggested in the reports by the OTC and NRC. In any case, the resolution of the genetic map had almost doubled, from the ten centimorgans of the White and Donis-Keller maps in 1987 to about six centimorgans. However, Jordan agreed that the currrent approach being taken by mappers—focusing on chromosomal regions which were already known to contain interesting genes—would likely not lead to a one-centimorgan resolution genetic map.

Olson then suggested that NIH's genome center simply own up to that fact and commit itself to a more realistic goal. Perhaps, he said, a five-centimorgan or two-centimorgan map would be a better target to shoot for over a five-year period.

The mostly likely cause of the slowdown was understandable. Churning out thousands of genetic markers and mapping them to chromsomes is a thankless, boring job, especially with no immediate likelihood of hitting paydirt. As Maynard Olson pointed out at the December meeting, it is difficult to get researchers interested in doing anything but local, high-resolution mapping, the kind of work that can quickly produce disease genes and a sense of accomplishment.

Ray White disagreed with that assessment, pointing out that *he* was still busy pursuing a map with one-centimorgan resolution. "Remember, a one-centimorgan map gave us the cystic fibrosis gene," he told *Science* magazine. He was referring to the high-resolution map of a part of chromosome 7 which he and others had prepared, and which was used to make the CF gene discovery. However, it was also true that White and his colleagues were focusing their mapping efforts on three chromosomes (5, 16, and 17) which happen to contain disease genes White is looking for.

Helen Donis-Keller (who by then had left Collaborative Research for a position at Washington University) placed the blame for the mapping slowdown on NIH. The real problem was an atmosphere of tight money which was making it difficult for both her and others to obtain funds for their projects. Producing lots of linkage markers and matching them to chromosomes is not the kind of work that excites peer reviewers of grant proposals, she noted. They tended to complain that such proposals "were not innovative" enough. That might be true, said Donis-Keller, but the point was not so much to be innovative as to get the job done.

The advisory committee finally decided to go along with Olson's suggestion to revise the goals of the Genome Project. The "medium-term" goal would remain a one-centimorgan map. The five-year goal would be a genetic map with an average resolution of two centimorgans, with no gap in the map greater than five centimorgans. At the same time, NIH's genome center began revising its strategy. A working group was set up to find ways of producing the one-centimorgan map and encouraging people to take part in the effort. One suggestion, said Jordan, would be to recruit people to work on mapping for short periods of time, and then let them make use of any data they produced. Until gene-mapping could be done entirely by machine, such inducements would be needed to entice researchers into doing the trenchwork necessary to create a high-resolution genetic map.

9

~~~~~~

Of Ethics and Trust

Know thyself.

Inscription at the Delphic Oracle

MANY PEOPLE find the very concept of sequencing the human genome to be ethically repugnant. This position is a variant on the idea that "there are some things that man [sic] was not meant to know." Some knowledge is by its very nature immoral or evil, and the possession of such knowledge is also immoral. Of course, the definition of "things that man was not meant to know" has changed over time. Only 450 years ago, for example, it was considered a serious sin to dissect human cadavers. The knowledge of the inner workings of the human body was forbidden. When the Flemish anatomist Andreas Vesalius published his illustrated book *De Humani Corporis Fabrica* in 1543 he violated that then-absolute tenet. Ninety years later Galileo violated another "not-meant-to-know" rule, using the newly invented telescope to examine the heavens and prove that the Earth was not at the center of the universe. He was brought before the Inquisition in Rome for doing so. More recently, many people have denounced the knowledge of nuclear fission as inherently immoral. The same damnation is now hurled by some at the science of genetics. For this reason, it is worthwhile to note the statement of ethicist Thomas H. Murray of Case Western Reserve University, quoted in the OTA report *Mapping Our Genes*: "The moral significance of humankind is no more threatened by peeking at the underlying musical notation, the base sequences, than is reading the score of Beethoven's last symphony diminishing to that piece of work."

In truth, the notion that genetic knowledge itself is "forbidden fruit" is not widespread, even among the most rabid of techno-fundamentalists. The majority of the Genome Project's critics would agree with Murray. What some object to is not so much reading the score as *rewriting* it. In other words, there is nothing immoral about merely sequencing the human genome. The ethical ambiguities arise when geneticists, molecular biologists, or doctors begin trying to change parts of the genome by means of genetic engineering technology. Related to this are the ethical and moral questions surrounding the termination of pregnancies—abortions—based on genetic knowledge of the embryo. A further extension of the moral and ethical questions occurs with the marriage of genetic information and computers. This moves the whole question of genomic ethics into the realm of the ethics of privacy and personal control of one's life.

However, there can be little doubt that the Genome Project will have enormous beneficial effects for the health of many people. Those curative accomplishments must be considered in the debate over the ethics of mapping the genome. One such important breakthrough was the discovery of the gene for cystic fibrosis. But there have been others, as well.

CANCER, MS, AND RP

One of the important clinical spinoffs of the Genome Project will be in the field of cancer diagnosis and cure. For example, mapping the human genome will lead to the identification of new *oncogenes*—genes that code for a predisposition for some form of cancer. The existence of oncogenes is indisputable. Their role in the development of several kinds of cancers is not in doubt. How many there are, and exactly how each of them work, is something that the Genome Project's accumulation of data will help answer. One good example has to do with one of the most common forms of cancer in North America: colon cancer.

In the United States and Canada, the lifetime risk of contracting cancer of the colon is close to 5 percent. One stage in the development of colon cancer is the development in the colon of fleshy growths called *adenomatous polyps.* The cancerous tumor itself is believed to develop within such growths. Colon cancer itself is preceded by several well-defined and identified genetic predispositions. One is the appearance of many such polyps in a person's intestinal tract. Doctors call this syndrome *familial polyposis coli,* or FPC. Another similar inherited syndrome is called Gardner syndrome, or GS. It is very difficult to tell the difference between the two, and together they are referred to as *familial adenomatous polyposis,* or FAP. The two syndromes are characterized by the growth of hundreds of these polyps in a person's colon. They usually appear by the time a person is 30 years old. If the polyps are not removed, the person runs a very high risk of developing colorectal cancer by age 40. Familial polyposis coli (and Gardner syndrome) are genetically dominant, not recessive, and they are autosomal. That is, the pattern of inheritance is one that indicates that the genetic cause is not tied to either the X or Y sex chromosomes, but to one of the other 21 autosomal chromosomes. The question is, which one?

In 1986 several researchers found out. They examined the chromosomes of a patient with both FAP and a large tumor. The researchers discovered that part of the long arm of the patient's chromosome 5 was missing. The missing regions were either 5q13-15 or 5q15-22. The designation is the standard "mapping language" for cytological genetic maps. In this case, "5" refers to chromosome 5; "q" refers to the long arm of the chromosome (if it were the short arm, the letter used would be "p"); and the "13-15" and "15-22" designate particular regions of that arm. Despite the absence of one or both of these parts of the chromosome, the patient still developed FAP.

Ray White, Jean-Marc Lalouel, and their colleagues at the Howard Hughes Medical Institute at the University of Utah Medical Center decided to try to narrow down the area in which the FAP gene might be located. They found it and published their results in the December 4, 1987 issue of *Science* magazine. They began their search by using a computer program to put together a genetic map of chromosome 5 with 16 DNA markers along its length. Then they tested several of the genetic markers to see if they showed up in the genetic material of people with FAP. Three of the RFLPs they were using turned out to be linked to the presence of familial polyposis. One marker in particular had a very strong linkage. The group theorized that it might even contain the FAP gene. It is called C11p11, and its location on chromosome 5 is around the area marked as 5q21-22 on a cytogenetic map of the chromosome. Somewhere in there, White's group predicted, is the gene or genes responsible for one of the predisposition signs of colorectal cancer. They hadn't yet found the gene itself, but they had considerably narrowed the area to be explored and mapped in search of it. It would now be possible for them or others to find new DNA markers for that particular region and map it in more detail. The step after that would be to clone different segments of the region and identify genes that produced messenger RNA found in colorectal cancerous polyps. Any gene that did that would be a prime candidate for the actual polyposi gene.

A few months later the White team had completed the next step in the search for the FAP gene. They published another paper, this time in the *American Journal of Human Genetics*. The researchers had continued their effort to find more genetic markers for the area on chromosome 5. They found six new ones in the FAP region of chromosome 5. The group then used a computerized genetic linkage program to analyze the appearance of seven genetic markers (the six new ones plus the C11p11 marker they had used earlier) in a group of 59 three-generation families. With the resulting data they constructed a genetic linkage map. White and his colleagues first determined that the gene or genes for FAP were actually about 17 centimorgans away from the C11p11 genetic marker, located at 5q21-22 on chromosome 5, which they had identified and spoken of in their December 1987 paper. Seventeen centimorgans are roughly 20 million base pairs. More importantly, the computerized genetic linkage analysis showed that the FAP gene location was actually very close to one of the new RFLP markers, one named YN5.48. The odds were more than forty thousand to one that the familial polyposis gene or genes were

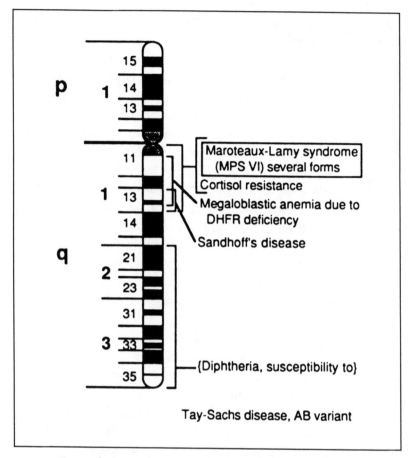

Researchers now have evidence for a gene on chromosome 5 (5q21-q22) that codes for a high predisposition to colorectal cancer. —Illustration reproduced from "Morbid Anatomy of the Human Genome," Victor A. McKusick (1988, Howard Hughes Medical Institute.)

near this new genetic marker, rather than near C11p11. White and his team estimated that the FAP genes are less than 2.5 centimorgans, or three million base pairs, away from the YN5.48 marker. That may seem a pretty far distance in ab-solute terms. However, it is a very small area when compared to the estimated 200 million-plus base pairs of chromosome 5 itself. The location of the FAP gene(s) has now been narrowed to an area comprising perhaps less than 1.5 percent of the area of chromosome 5.

With this discovery, Ray White's group had opened the way for the development of practical preclinical diagnostic tests for FAP. Colon cancer finally appears only after several genetic events take place. For example, the mutation of part of chromosome 12 takes place in more than a third of all cases of colon cancer or cancerous polyps. Other genetic changes appear to take place on chromosomes 17 and 22. A FAP screening test would look for the presence of the YN5.48 marker on chromosome 5, since the FAP gene or genes are extremely close to that marker and are almost never cut off from it when chromosomes recombine during the fertilization of an ovum. Such a test would then be the "gateway" to other tests for other genetic predispositions, such as the mutations and changes connected to colorectal cancer which are found in chromosomes 12, 17, and 22. If negative, these tests—which are not far in the future at all—would provide a great deal of reassurance for members of families with a history of colorectal cancer. Positive tests would quickly identify those people who have a high *predisposition* for colorectal cancer, not the disease itself. And the identification would take place years, even decades, before the disease would be likely to strike. Such genetically predisposed people would therefore be able to make changes in diet and lifestyle that would greatly reduce their chances of getting colon cancer.

One of the more exciting genetic breakthroughs in 1989 was the discovery of a gene that codes for a much higher risk of multiple sclerosis, or MS. The research leading to the new discovery was done by a team that included Stephen Hauser of Massachusetts General Hospital in Boston.

Multiple sclerosis destroys the fatty tissue called myelin which surrounds parts of the body's nerve cells. Myelin acts like insulation. When the sheathing is destroyed, nerve transmission is disrupted, causing symptoms that range from tremors, muscular weakness, paralysis, and impaired vision.

The exact cause of the myelin destruction is still unknown, but most researchers now believe that MS is an autoimmune disease. For some reason, the immune system's T-cells become misguided and begin attacking the body itself, in this case the myelin sheath.

Researchers have long suspected a genetic link to MS. For example, a person whose identical twin has MS is much more likely to come down with the disease than another sibling would be. Also, some scientists have recently found certain forms of the gene coding for the T-cell to be associated with MS in individual patients. However, no one had found any inheritance pattern in families for any MS-associated T-cell gene. In fact, no one had yet looked at the T-cell gene in its entirety.

Hauser and his team knew about the T-cell hypothesis. They decided to look for mutations in a gene known to code for the protein receptor (a molecular "lock" as it were) on the surface of the T-cell, which it uses to recognize its alien targets. If the T-cell has an abnormal receptor, it would be more likely to malfunction, attacking the body's own cells and perhaps causing MS. By using a modified form of pulse-field gel electrophoresis, the researchers were able to view the entire unbroken gene and make comparisons between genes from different people. They examined the two copies of the T-cell receptor gene from 40 pairs of siblings who all had multiple sclerosis. The researchers confirmed, first of all, that there was a difference between the normal form of the gene, and the gene in MS siblings. If the genes Hauser was looking at had not been involved with MS, their distribution would be random. Ten of the sibling pairs would have the same two gene copies from their parents, 20 would have one of the gene copies in common, and ten would have neither gene in common. The distribution was far from random, Hauser and his colleagues found. *Fifteen* of the sibling pairs had inherited the same two forms of the T-cell receptor gene from their parents. Twenty-two of the sibling pairs had one of the gene copies in common. Only three of the sibling pairs shared neither gene copy. The research by Hauser and his colleagues also revealed that people carrying this particular form of the T-cell receptor gene are more than three times as likely to contract MS as the general population. Siblings of MS patients have 20 times the risk of the general population, so Hauser's findings suggest that several different genes, *as well as environmental factors*, are involved in getting MS.

Another major breakthrough in 1989 was the discovery of a gene that causes one form of the eye disease *retinitis pigmentosa*, or RP, which afflicts more than 100,000 Americans. RP is characterized by the degeneration of cells in the retina, the layer at the back of the eye that detects images and sends them through the optic nerve to the vision center at the back of the brain. It progresses from night blindness to tunnel vision, and finally to complete blindness. RP comes in several forms. In one of them each child of an affected parent runs a 50 percent chance of getting the disease. This form makes up about 20 percent of all RP cases. About half of them, in turn, may be caused by the newly-discovered gene. A gene which causes another type of RP was found several years earlier, but has not yet been isolated.

The discovery of the new RP gene was made by a group of researchers led by Peter Humphries of Trinity College in Dublin, Ireland, and Stephen Daiger of the University of Texas Health Science Center in Houston. The researchers studied a family near Dublin which has a hundred living members—a very large pedigree, as geneticists would say. Half of the members have RP. Humphries, Daiger, and their colleagues compared the DNA of affected and unaffected members. In particular, they traced the inheritance patterns of RFLP markers on the chromosomes. Their analysis of the RFLP inheritance patterns found a marker lying extremely close to a gene on chromosome 3. The researchers believe that this gene is responsible for this particular form of retinitis pigmentosa.

The discovery of the new RP gene using RFLP mapping and linkage analysis technology is the first step to isolating the gene and then determining which protein or proteins it makes. That will reveal the actual mechanism by which the gene causes its form of RP. And *that*, in turn, will eventually lead to diagnostic tests and perhaps even genetic cures for one form of retinitis pigmentosa.

* * *

THAT THE GENOME PROJECT will have an effect on the questions of medicine and ethics cannot be denied. The question is how profound will that impact be. Ray White, one of the leaders of the Genome Project, has not avoided thinking about those issues. However, White feels very strongly that one of the most important effects of the effort to map the human genome will be to bring the *facts* to bear on the debate over ethics. White also believes there is an overemphasis on the role of genetics in the development of the human individual. Most people think that genetics play more of a role than is really the case. The specific complement of genes that one has does play a significant role in making each of us who we are, of course. But it is only one component of what is mixed into the pot called the individual. The genetic hypothesis, in White's opinion, today obscures many other components, including environment and childhood upbringing.

THE "OLD-FASHIONED WAY"

White also notes that we are asking the same ethical problems today about the Genome Project that were asked nearly two decades ago, when genetic engineering became a reality. The main ethical issues were, in fact, exhaustively addressed in a 1973 book entitled *Ethical Issues in Human Genetics: Genetic Counseling and the Use of Genetic Knowledge*, edited by Bruce Hilton and four other ethicists. This was a scholarly volume of proceedings from a conference on ethical issues in genetics which took place in 1971. The first essay in *Ethical Issues in Human*

Genetics, by Tracey Sonneborn, laid out three general classes of ethical and moral problems.

The first major set of ethical issues centers on the possibility of human procreation by means other than normal sexual intercourse. In 1971 the only such form of procreation available to humans was artificial insemination. Even then, however, ethicists and others were aware of animal research which would eventually lead to *in vitro* fertilization. The first known case of successful human *in vitro* fertilization took place in 1978. The fertilized egg was reimplanted in the biological mother's womb and carried to term. Louise Brown is today a healthy and normal young girl. Ethicists were also concerned about the issue of surrogate mothers. That issue, too, has become a reality. It has given rise to a whole new set of court law. In fact, one can plausibly imagine a human child having five "parents": the biological mother; the biological father; a surrogate mother who carries the fetus to term in her womb; and a married man and woman who legally adopt the newborn child. It is legal in some states for gay men and lesbians to adopt children, so one can imagine a similar scenario in which the child has three "fathers" and two "mothers," or one "father" and four "mothers." The Genome Project will probably have no direct bearing on these and other ethical issues.

BABY SHOPPING

It will, however, have an impact on a second major set of ethical issues addressed by *Ethical Issues in Human Genetics*: the then-emerging practice of genetic counseling based on diagnostic tests which could detect genetic abnormalities in developing human fetuses. Foremost among these is whether it is ethical to abort a fetus found to have abnormal or defective chromosomes or genes. Abortion, and a woman's right to choose whether or not to undergo this medical procedure, has become the preeminent popular moral issue of the moment. The fundamentalist Christian churches, along with the Roman

Catholic Church, have taken the position that abortion is immoral under nearly all circumstances. Some extreme proponents of this position insist that abortion is immoral under all circumstances, including cases of rape, incest, or the danger of death to the mother. This position is based on the religious belief that a fertilized human ovum is a human being in every explicit and implicit sense. To abort a human fetus or embryo at any stage of development is, under this position, tantamount to murder. For this reason, most of these churches are either opposed in principle to the use of genetic testing *in utero*, or to the use of such tests to decide whether or not to have an abortion.

This set of ethical conundrums, and the religious opposition associated with it, has existed for many years. The Genome Project will not increase the information level of the debate. Definitions of "human" which focus on the existence of an immaterial soul or spirit are basically religious and spiritual in nature, not medical or scientific. The Project will likely increase only the debate's noise level. As more and more genes responsible for genetic abnormalities and diseases are discovered and mapped, new diagnostic tests will be developed and used by couples who are newly pregnant. More and more people will thus be faced with serious ethical questions: Do I abort a fetus known to have a serious genetic abnormality? How "serious" is "serious"? If I do not, and the child is born and grows up deformed and resents me, will the child be able to sue me for "wrongful life"? Doctors, too, will have some serious questions to answer: If a doctor does not offer genetic screening as an option to a pregnant woman, and the woman bears a genetically defective child, can the doctor be sued for malpractice? If a woman refuses to have a genetic screening test done on the fetus, and the child is born with a genetic disease, can the doctor *still* be sued for malpractice? Can the *child* slap the *doctor* with a "wrongful life" suit? Can a doctor ethically refuse to treat a pregnant woman who refuses to have a genetic screening test? Then there are the legal complications arising from these questions—not to mention the insurance companies weighing in with their opinions.

"O BRAVE NEW WORLD!"

The third major set of ethical issues centers around what Sonneborn called "genetic surgery" and what White has called "gene implantation." This area of genetic technology is still in the realm of science fiction. However, it won't be for much longer. Genetic surgery or genetic manipulation is the process of changing the genetic complement of a human being by either cutting certain genes out of a chromosome or implanting genes into one. There are two major aspects to this—somatic cell manipulation and germ cell manipulation. To change the genetic code of somatic cells—the cells of the liver, kidneys, bones, brain, skin, and so forth—is to change the genetic code of *that one specific person*. Ray White does not believe that somatic cell gene implantation will be a significant way in which Genome Project information will be used. More likely, says White, will be the development of drugs based on knowledge of what genes are doing and what the biological systems are about. These new drugs, offspring of the Genome Project, will either cure or control genetic diseases by turning off mutated genes or turning on other genes. One can imagine, for example, a drug that switches on an anti-oncogene, a gene whose protein product acts to prevent a certain kind of cancer from developing. Such drugs will be the future cures for different cancers. In any case, almost no one sees any ethical problems with genetic manipulation or surgery on somatic cells.

The serious ethical questions arise with the issue of germ cell manipulation. One in particular is of great concern to Ray White. He, along with nearly everyone else in biology, feels that genetic manipulation of human germ cells (eggs and sperm) violates the primary canon of human experimentation—the consent of the subject. The individual who agrees to have his or her germ cells changed can consent. *But that person's progeny are now committed to an experiment to which they did not consent.*

However, White in particular does not see this as an absolute barrier to germline genetic manipulation. Our knowledge and understanding of our genetic information will eventually reach a point at which germ cell manipulation will not be experiment, but therapy. When that point is reached, the concern about the consent of the experimental subject will disappear. The problem, of course, is that science and medicine still have to get from here to there. And Ray White freely admits that he doesn't know how that will happen.

GENOMIC ETHICS AND THE OTA REPORT

The OTA report's summary of the ethical problems raised or highlighted by the Genome Project is cogent. Many of its points are those mentioned by Sonneborn, with the advantage of an additional sixteen years of debate. Some are new. Some are uniquely related to the Genome Project itself.

One of the reasons for studying the human genome, as well as other genomes, is to see how individual genetic variations create (or have an effect on) differences among individuals. What will happen when we learn these answers? What will be the impact upon our society and culture—on job security, insurance premiums, civil rights, education, religion—if we learn that genetic variations play a larger role than previously suspected? Or a smaller role? The OTA report points out that debates about ethics, about "what *ought* to be done often cannot be resolved by empirical inquiry. Specific genetic information such as the location of a gene along a chromosome or the sequence of nucleotide bases composing a specific gene is value-neutral." These are not really ethical questions. However, questions involving the availability of genetic information about an individual or the entire human species, or what constitutes "proper use" *are* ethical questions. They involve choices to be made by fallible humans and based on different and conflicting notions of what is good or evil, right or wrong, desirable or obscene.

The Genome Project, for example, is again focusing attention on ethical questions surrounding the very conduct of basic scientific research. The OTA report asks

■ "How should the conduct of research in the basic sciences, such as genome mapping and sequencing, be influenced by a concern for the social good?

■ "What are the considerations when basic research in the biological sciences seems to take resources away from areas of research that might have more immediate social benefit?"

The first question continues to be fiercely debated. It is a question asked constantly, if often only implicitly, in Congress and in the Oval Office. Will research in this scientific field be likely to yield results that will be good and desirable for society at large, or for significant portions of society? When applied to *basic research*, this question is fundamentally meaningless. There is no way to predict what the practical results of *basic research* will be. For example, in 1905 no one, not even Einstein himself, could have foreseen the practical consequences of his explanation of the photoelectric effect. Yet today, a practical consequence of that basic mathematical research is one of the most powerful social forces on the planet: the television industry. The scientific community has long argued, with considerable merit, that basic scientific research is valuable by itself, and that it must be pursued for its own sake.

However, the question does have meaning when applied to the tools that result from basic research, to the technologies that evolve. The Genome Project is not in itself basic research; it is a technological application of basic scientific breakthroughs made by Watson, Crick, Berg, Gilbert, Maxam, and others. In this case, an argument can be made that a massive scientific and technological project like the Genome Project, which will involve great amounts of funds and time, should be subject to "a concern for the social good."

The second question is one that has been addressed by Genome Project proponents almost from the beginning. Researchers like David Baltimore legitimately raised the question of limited monetary resources. Would it not be better—that is, more ethical—said Baltimore, to spend our limited resources on fighting AIDS rather than on pursuing a Big Biology project of mapping the genome? The scientific community, along with politicians in Congress and the White House (the sources of the money), eventually decided that a genome project would not adversely affect other areas of biological scientific research. Furthermore, they agreed that the Genome Project was itself a scientific endeavor which would have an overall beneficial effect on society.

GENOMIC ETHICS IN THE INFORMATION AGE

Several ethical issues raised in the OTA report are unique to our Information Age. They are related to the issues surrounding genetic counseling, but were not addressed specifically in *Ethical Issues in Human Genetics*. The report asks, among other things,

■ Who should have access to map and sequence information in data banks?

■ What are the ethical considerations pertaining to control of knowledge and access to information generated by mapping and sequencing efforts?

■ Do property rights to individuals'genetic identities adhere to them or to the human species?

■ Who owns genetic information?

It may be argued that human genetic information by its very nature lies in the public domain. No one may copyright or patent something which is part of the essence of being human, which is held in common by all who partake of the human heritage. Current patent law, for example, does not allow the patenting of a person or an idea. An opposing view holds that genomic information is not commonly knowable (how many people do you know who understand how to read a contig map?) and requires expensive machinery to read. Therefore, it is reasonable that the fruits (economic and otherwise) that come from discovering genetic sequences should go to those who do the work. This is "The laborers deserve their wages" position. It does not matter whether or not the genetic sequence is unique or "the common heritage of humanity." It does not even matter how it is used. What matters is that it is valid intellectual property, made so by the effort and inventiveness of the one who discovered it. Current patent law also recognizes this argument.

Several of these questions arise from the computerization of our society and the flow of information. It is quite likely that, within the next 50 years, all people will have complete maps of their specific genetic makeup filed in a computer database at birth. How secure will that data be? Who will be able to read it? Just you? Your spouse or partner? Your insurance companies? Your employer? The federal/state/regional/local government? There could easily be serious problems related to insurance liability, or to getting a job, for those whose "computerized genome readout" shows them to be at risk of some genetic disease. Ray White likes to point out that we are all at risk of something. Everyone—doctors, lawyers, insurance executives, politicians—will die of *something*. Thus, the discriminatory power of future genetic tests will eventually diminish.

Rather, people will be able to make decisions about their lives based on this detailed information about their genetic makeup. Doctors, in turn, will be able to cure acute medical conditions more successfully. For example, individual genome data may be imprinted on a small card carried by every person. When this person goes to the doctor's office for a regular checkup, the doctor will take the card and run it through a magnetic reader and get an instant readout of the person's genome. The doctor will be able to take that information and mesh it electronically with the person's current medical status, dietary plan, workplace environment, and daily exercise plan. The result will be a thorough understanding of the person's current health and likely future trends, along with expert advice on how to stay healthy. A genome card will thus make possible the ultimate in high-tech preventive medicine. To many people this would be a dream come true.

To Jeremy Rifkin, it is likely a nightmare.

THE RIFKIN SYNDROME

Jeremy Rifkin is the gadfly of the biotechnology revolution. Since 1977 and his book *Who Should Play God?*, this veteran of the Vietnam antiwar movement and organizer of the 1976 People's Bicentennial Commission has been incessantly calling for a society-wide examination of the ethics and morality of all phases of biotechnology. His supporters see him as the leading voice for prudence and caution. His opponents consider him little more than a Luddite demagogue interested only in destroying a field of science. In truth, Rifkin is somewhere in between.

Despite the "Luddite" charge, Rifkin and his activist colleagues have spearheaded some beneficial activities. For example, they forced the Pentagon to abandon plans for a biological warfare lab in Utah; exposed some significant shortcomings in the government's regulations of the release of genetically altered organisms into the environment; and championed improvements in the storage of botanic germplasm, those seeds which provide the genetic base to the world's food supply.

On the other hand, Rifkin has often supported actions which were often dubious and at times downright silly. His series of lawsuits and administrative maneuverings delayed for four years the testing of Frostban, a genetically engineered microbe which protects plants from frost. David Baltimore called that campaign "absolutely preposterous," since the Frostban microbe is actually found in nature. Rifkin has also opposed the splicing of human genes into laboratory animals, calling it an immoral breaching of the "species integrity" of animals. In fact, Nature herself has been "breaching" the "integrity of species" for more than four billion years: it's called evolution. Furthermore, the genetically related diseases that researchers study in animals—arthritis, cancer, diabetes, heart disease—occur naturally in the species being studied.

Speaking of evolution, Rifkin has written that the Darwinian theory is dying. In fact, says evolutionary expert and award-winning author/scientist Stephen Jay Gould, it has never been healthier. Rifkin's evolutionary *faux pas* is an example of one of his greater weaknesses, a preference for the grand, sweeping statement. In one case, he called animal gene-splicing "the greatest assault on animal welfare in history," which will turn them into products "indistinguishable from microwave ovens and automobiles." On another occasion, he claimed that the "bankrupt idea of 'progress' has led the entire planet, the whole solar system, to where it's reeling." Those phrases may well sound catchy to newspaper headline writers, network news anchors and much of the general public, but they are far from accurate depictions of the status of biotechnology today—or any time in the near future. And while the idea of "inevitable scientific progress" may well be bankrupt, it is hardly causing the entire solar system to "reel."

However, Rifkin's greatest weakness has often been in the area of scientific facts. It is not that he is stupid. He is not. One biologist has been quoted as saying that Rifkin "is extremely intelligent, and grasps scientific issues immediately, if you take the trouble to explain them." But as his erroneous comments about evolution illustrate, he sometimes does not take the time to understand them. While many reviewers have praised his books *Algeny* and *Entropy* as exciting and well written, most scientists were horrified by them. Gould called *Algeny*, for example, "a cleverly constructed tract of anti-intellectual propaganda masquerading as scholarship." His writing about science is rich with allusion and metaphor, and spiced with entertaining (and often terrifying) anecdotes. They are also weak on accurate explanations of scientific fact and display a lack of understanding of how science works and scientists behave. Rifkin's goal is to raise "big questions" about ethics and morality, and to challenge the entire scientific paradigm as he perceives it. He only damages his cause with his consistent display of scientific semi-literacy.

For all his huffing and puffing about genetic engineering and the biotechnology industry, Rifkin remains curiously ambivalent about the Genome Project. According to a 1988 article in the *New York Times Magazine*, he does not oppose the Project per se. He does insist that citizens play a role in puzzling out the ethical quandaries highlighted by the drive to map the genome. It's a reasonable request, and one with which at least one scientist involved in the Genome Project would agree. Leroy Hood has said that the ethical questions connected with genetic engineering and gene-mapping are not issues that only scientists must solve. Everyone in our society, he has insisted, must grapple with them and contribute to the debate. For the medical and social consequences of the Genome Project will affect everyone.

Rifkin has also said that he has no quarrel with recent efforts at gene therapy, the attempt to cure a genetic disease by either eliminating the defective gene or replacing it with a properly functioning one. At the same time, however, he has claimed that these pioneering trials are examples of science focusing on technology without any thought to its ethical implications. And *that* is something with which many, if not most, genetic researchers would disagree. Almost without exception, they have thought about the ethical and moral implications of implanting human genes in animals for research purposes, and both human and animal genes in humans for medical purposes. When one is faced with a fatal genetic disease, and one has the technological tools to cure that illess, it is immoral *not* to use those tools.

The same argument applies to work on the Genome Project. "If you want to understand Alzheimer's disease," James Watson has said, "then I'd say you better sequence chromosome 21 as fast as possible. And it's unethical and irresponsible *not* to do it as fast as possible."

Rifkin and his supporters charge that the rights of animals used in genetic research—including, presumably, the Genome Project—are being violated. Also warning that the risks in gene therapy may be too high, they often resort to genetic Frankenstein metaphors. A general public familiar with grade-B science fiction movies, unfamiliar with risk assessment, and having little scientific education finds such statements scary and believable. In point of fact, there has never been a release of gene-engineered malevolent organisms into the environment. Rifkin's concern about risks to people taking part in gene therapy experiments also rests on shaky ground. Each person who takes part in such a trial must first, by law, have all the risks and benefits explained, and must voluntarily sign a release stating that fact. Most such patients know their disease is fatal; they know that other treatments have failed or don't even exist; as individuals they have nothing to lose and possibly a life to gain; and their participation in such experimental trials will contribute to the eventual health and well-being of many others.

Finally, Rifkin's concern for the rights of research animals is laudable. In a culture as deeply estranged from Nature as ours, it is not surprising that animals are usually treated as objects to be used and thrown away. It is not an attitude confined to laboratory animals; ask anyone who has worked in animal shelters or with the SPCA. In a culture which experienced and celebrated the intimate connection between humans and all other living beings (perhaps grounded in some version of Buddhism, Native American spirituality, or the ancient European nature religions?), research animals would certainly be treated differently than they are today. However, attacking genetic research on the basis of animal rights violations will not bring on the New Age Millenium, much less convert scientists into animal rights activists. The upshot has been that Rifkin has—perhaps unfairly—been seen as being more interested in the welfare of nude rats than in a woman who has the genes for Huntington's disease.

WATSON AND GENOMIC ETHICS

Ethical concerns are certainly not at the top of the worry list for most genomic mappers and researchers. At the Human Genome I Conference in San Diego in October 1989, only 30 minutes were devoted to a formal discussion of the social and ethical implications of the Genome Project. Despite that, it is important to remember that Jeremy Rifkin does not have a monopoly on questions about the ethical issues of biotechnology and gene mapping. Nearly all of the major players in the Genome Project—including Lee Hood, Victor McKusick, Ray White, and James Watson—have publicly and privately voiced their concern about such issues. In fact, Watson's been pondering them for at least 20 years. In an article he wrote for *Atlantic* magazine in 1971, Watson said: "If we do not think about [the ethics of genetic technology] now, the possibility of our having a free choice will one day suddenly be gone." Now, in his role as overseer of the NIH end of the Genome Project, Watson is taking an aggressive position on ethical discussions. At his insistence, three percent of NIH's Genome budget will be devoted to studying the ethical implications of mapping the human genome. That's about $6 million a year out of an annual budget of $200 million, and close to $90 million over the proposed 15-year span of the Project. That is undoubtedly the most money ever spent specifically on biomedical ethics—or perhaps on any kind of ethics. Even Jeremy Rifkin must be moderately impressed by Watson's move.

NIH also has an ethics working group for the Genome Project. Its mission is to anticipate any abuses which might occur using knowledge attained by the Project and prevent those abuses before they take place. To that end Watson has deliberately packed the group with people attuned to the misuses of scientific knowledge. The group includes bioethicist Patricia King of Georgetown University; Howard University's Robert Murray, Jr., an expert on the abuses of genetic tests for sickle-cell anemia; and Jonathan Beckwith of Harvard Medical School, a frequent critic of genetic technologies. Chairing the ethics working group is Nancy Wexler of Columbia University. Wexler is a neuropsychologist who has been a leader in finding the genes for Huntington's disease. The ethics of genetic screening tests and how they will be impacted by the Genome Project are of great personal interest to her. The Huntington's gene runs in her family, and Wexler's own mother died of the disease. She has chosen not to test herself for the presence of the gene. "The fact that Watson opens his mouth and utters the word 'ethics' changes the whole valence" of the discussion, Wexler has been quoted as saying. "Other people start to think, 'Gee, this must be important. This is something we ought to pay attention to.' "

What Watson told his fellow scientists at the Human Genome I conference summed up his position:

It would be naive to say that any of these answers are going to be simple. About all we can do is stimulate the discussion, and essentially lead the discussion instead of having it forced on us by people who say, 'You don't know what you're doing.' We have to be aware of the really terrible past of eugenics, where incomplete knowledge was used in a very cavalier and rather awful way, both here in the United States and in Germany. We have to reassure people that their own DNA is private and that no one else can get at it. We're going to have to pass laws to reassure them. [And] we don't want people rushing and passing laws without a lot of serious discussion first.

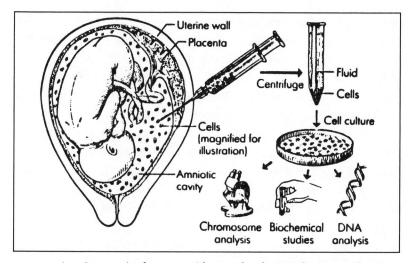

Amniocentesis, the most widespread technique for prenatal genetic diagnosis and genetic counseling.—Reprinted from Pines, Maya, "The New Human Genetics," NIH Publication 84-662 (U.S. Department of Health and Human Services. National Institute of Health. National Institute of General Medical Science, September 1984.)

BOY OR GIRL? LIFE OR DEATH?

It's not that Watson thinks there will be *no* misuse of information gleaned from the Genome Project. Humans are human. Some will misuse the new genetic knowledge unwittingly; others out of genuine compassion; still others in the service of racist or other immoral ideas. It's already happening, in fact, and one good example is the sex selection of babies.

In India and Asia it is already becoming common for pregnant women to seek tests of the fetus to determine its sex. Female fetuses are frequently aborted. In these cultures, girl children are much less valuable than boy children. Girls and women are in fact considered not much more valuable than many animals—certainly less valuable than a healthy water buffalo. Female infanticide has long been common, and aborting a female fetus little more than a temporal extension of this practice. Sex-determination fetal tests are also done in

America, but for different reasons. The same sexist attitudes exist here, but Americans consider infanticide morally abhorrent. Ethical and moral debates over abortion aside, no reputable doctor would perform an abortion if he or she knew it was simply to get rid of a female fetus. Indeed, women's health clinics would be outraged at such a suggestion. Most genetic counselors would strongly object to having their services used for sex selection. When American couples practice sex selection, it is often to avoid passing on sex-linked genetic diseases, such as certain forms of muscular dystrophy. Genetic counselors and obstetricians make a resolute distinction between selection to avoid genetic diseases, and selection for particular genetic characteristics such as sex or eye color.

Should there be some kind of law against genetic counseling or tests for sex selection? Currently the only movement in this regard seem to be those linked to restrictive anti-choice legislation in various states. Most pro-choice advocates agree that abortion for sex selection is morally wrong, but resolutely oppose the repressive bills which include such bans. A law which addressed only the ethical and moral issue of abortion for sex selection might stand a better chance of passing in various state legislatures. However, it would still not address the issue of genetic counseling or testing for sex-linked traits, or any other genetic conditions. Anti-choice activists oppose abortion for nearly any reason; some would ban abortions even in the case of rape, incest, or the sure threat of death to the mother. Many also oppose genetic counseling and the development of genetic testing, since they feel that "people can't be trusted" to make "righteous Christian choices." Their attitude seems to be that it is better to force people to "be good" rather than allow them the exercise of free will and the likelihood of making a "sinful choice."

However, those with a more pro-choice view of both abortion and free will are still troubled by the ethical choices represented by genetic counseling—both for sex selection and for other reasons. Within 15 years the Genome Project will have

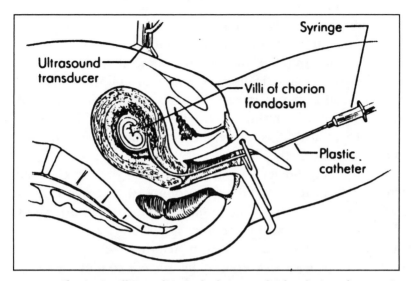

Chorionic villi sampling is the latest medical technique for genetic diagnosis of the fetus.—Reprinted from Pines, Maya, "The New Human Genetics," NIH Publication 84-662 (U.S. Department of Health and Human Services. National Institute of Health. National Institute of General Medical Science, September 1984.)

uncovered the genes which code for diseases like Alzheimer's, forms of cancer, heart disease, and manic depression—or at least for strong tendencies to contract them. What will people do with this knowledge once they have it? Is it morally acceptable to abort a fetus because the human it will become *might* someday be afflicted with insulin-dependent diabetes, or heart disease, or manic depression?

The question of "wrongful life" lawsuits also comes up. Several such suits have come to trial, with large monetary awards to the person "wrongfully" allowed to live. It is not inconceivable that a 25-year-old man born with the gene coding for Alzheimer's disease might hit his parents and the doctor or midwife who delivered him with a wrongful life suit. And it is not inconceivable that he would win. Should he? Should he even be allowed to file such a suit? Should such a lawsuit be allowed if the parents or doctor or both knew from genetic testing that he had the Alzheimer's gene?

And what will such a scenario do to the insurance industry? It's quite possible that (1) health insurance rates will skyrocket, and/or (2) insurance companies will refuse to insure for a long list of genetic diseases, and/or (3) millions of people will demand a change in such insurance practices, and/or (4) some enterprising businessmen will step in and create some new form of health insurance, and/or (5) the health insurance industry will collapse.

In the end, however, it is also possible the the Genome Project will solve at least the legal conundrums in one 15-year fell swoop. For mapping the human genome will clearly show that we are all equal in one very important way. *We are all genetically flawed.* It is already estimated that each of us carries a recessive gene for an average of five genetic conditions which would be fatal for children who inherited another copy of the same gene from the other parent. It is equally likely that we all carry genes which code for tendencies to genetic disorders that, like multiple sclerosis or heart disease, might be triggered by a virus or some environmental condition. As one medical ethicist has observed, the Genome Project may well teach us to value our blemishes. Perhaps it would be worthwhile for each of us to carry in our wallet or purse a small card which reads: "I'm not perfect. I'm only human," and read it frequently.

* * *

TRACY SONNEBORN HAS some wise words for all who contemplate the serious ethical issues that surround the entire field of genetics, including the Genome Project. At the end of his introductory essay in *Ethical Issues in Human Genetics* he writes:

> [We develop our] ethics by the method of public discussion, by individual decisions and actions, by public acceptance of what appears to be right and good for man [sic], and by rejection of what appears to be wrong or bad. We agree that it is right and good to reduce misery and improve the quality of life for all those who live, by using environmental and

social means. We now debate whether it is right and good to use genetic means. Our conceptions of what is ethical, right, and good change in the light of new knowledge and new conditions. What we lack is neither flexibility of mind nor adventurous spirits, but knowledge and experience. If the future can be judged by the present and the past, we shall get that knowledge and experience and eventually authorize the ethics that permits doing what is believed to be right and good for man [sic].

. . . . We have no basis for being cocky. We are still full of ignorance in spite of the spectacular increase of knowledge. It would be both unwise and foolish to proceed without the utmost humility and compassion.

In his "Physician's Oath" Hippocrates reputedly said: "I will use treatment to help the sick according to my ability and judgment, but never with a view to injury and wrongdoing." Ability and judgment must go hand in hand, with the aim being not to cause harm but to heal.

Today, medical science is presenting medical technology with the ability to do things once only dreamed of by science fiction writers. By combining the ever more sophisticated tools of genetic engineering with maps of the human genome, it will soon be possible to abolish some genetic diseases from the face of the earth. Anyone who has had a child die horribly of Tay-Sachs or Lesch-Nyhan; or a teenager afflicted with cystic fibrosis choke to death on his own mucus; or a parent or grandparent sink into the ignominious dissolution of Huntington's or Alzheimer's; all these people—and the vast majority of everyone else—will welcome to greater or lesser degree the possibility of curing such illnesses. Within five to ten years, and probably sooner rather than later, we will have the ability to cure one or more of these genetic diseases. The first steps are about to be taken. The first attempts at "gene therapy" are just around the corner.

A group of researchers is now planning to use genetically engineered viruses to cure a genetic disease called SCID, or

Severe Combined Immunodeficiency. This extremely rare condition (less than three dozen cases are known worldwide) results in the almost complete absence in a person of a functioning immune system. The experiment could begin some time in 1991, if all the regulatory requirements are met and there are no major lawsuits brought against the researchers to stop it. Should it prove successful, and any possible legal objections are taken care of, other experiments will surely follow. Scientists will continue to develop more sophisticated and reliable methods for gene therapy.

Will our judgment keep pace with our abilities? There is no certain answer to that question. Unlike the color of our eyes or the wave of our hair, good and wise judgment is not genetically inherited. It is an acquired characteristic. Nor is it something taught in schools—at least, not in most schools. The ability to make good judgments—to choose to use a new tool or medical technique or to choose *not* to use it; to say, "This particular use of gene therapy is ethically good. This reason for aborting a fetus is morally wrong"—is something each person *learns*. We learn it from our parents' example. We learn it from our own experiences and from our own choices. We learn it from the responses of others to our actions and choices, and from the still small voice of our own consciences to those choices.

In a real sense, the way we answer this question boils down to how we perceive human nature. People are either inherently bad, inherently good, or somewhere in between. Some of us consider our species as being inherently bad. Given a choice between doing something good and kind (they say) and something mean and cruel, and all other things being equal, people on the whole will choose the mean and cruel action. Others among us consider people as being basically good. Given the same kinds of choices and conditions, they say, people will tend to choose the good. Still others think we fall somewhere in between. Clearly, they note, some people deliberately choose to do evil. The Mengeles and Bundys among us are all too real. On the other hand, so are the Ghandis and Mother Teresas. People can and do make good moral and ethical judgments, and they do so usually with only incomplete and sometimes contradictory information. Such judgments are made "from the gut," with our intuitive sense. Those who are so distrustful of the public's ability to make good judgments on the matter of genomic mapping and genetic engineering may simply misunderstand the basic nature of such judgments. They are, in the end, not logical but intuitive. Perhaps, individually and collectively, we need to trust more in our intuition, supported with clearly presented information.

Finally, can we refrain from using this new medical technology for "injury and wrongdoing"? The pessimists among us will point to episodes both ancient and modern of germ warfare; to Auschwitz and the evil medical experiments of the Nazis; to the criminally tardy response to the AIDS epidemic. Here, they will say, is proof that "man cannot be trusted with such knowledge." The optimists will point to the work of Salk and Sabin in developing polio vaccines; to the global elimination, under the auspices of the World Health Organization, of smallpox; to the growing acceptance—even by some "orthodox" medical practitioners—of "unorthodox" healing practices such as therapeutic touch, acupuncture and acupressure, and other holistic methodologies. See, they will say, people *can* change, they *can* grow, they *can* learn new things (or relearn old things) and use them wisely and well.

It comes down to this: Trust in ourselves and trust in others. Do we trust in our ability to learn from our past, from our mistakes, from our successes as well as our failures? Do we trust in our ability to be humble, not proud, but connected to the earth, grounded?

Do we trust in our ability to grow?

Do we trust in others' ability to grow, to learn?

Mapping the human genome will be the greatest scientific and technical achievement of this century, greater even than the invention of the atomic bomb. We do not know the final shape of the future we are creating with the Genome Project. If the history of prognostication is any guide, it is likely that our future will be wildly different from our most fevered dreams.

Still, it is time, as we move into a world whose shape is still only dimly discerned on maps of chromosomes and cosmids, to embrace trust.

If we trust in ourselves and each other, and listen to ourselves and each other, and strive to make wise judgments, and accept the fact that we will make mistakes, and still strive to trust and listen and make choices; if we do these things, the journey we take using the new maps we are about to make will be an exciting one.

Selected Bibliography

The following list of books, articles, and technical papers related to the Genome Project is not meant to be exhaustive. It does include much of the material used during the research for this book, as well as other items of interest. Those who wish to read in more detail about some of the topics discussed in this book may wish to peruse these articles, technical reports, government documents, and books.

Popular Articles and Books

Angier, Natalie. 1986. A stupid cell with all the answers. *Discover*, November, 71–83.

———. 1989. The gene dream. *American Health*, March, 103–8.

Atwood, Cynthia. 1988. *Yale molecular biologist champions international database system*. New Haven: Yale University Office of Public Information.

Bankowski, Zbigniew. 1988. Genetics, medicine, and ethics. *World Health*, December, 3–5.

Barinaga, Marcia. 1987. Critics denounce first genome map as premature. *Nature*, 15 October, 571.

Barnes, Deborah M. 1988. Schizophrenia genetics a mixed bag. *Science*, 18 November, 1009.

———. 1989. Fragile x syndrome and its puzzling genetics. *Science*, 13 January, 171–72.

———. 1989. Troubles encountered in gene linkage land. *Science*, 20 January, 313–14.

Barnhart, Benjamin J. 1989. DOE human genome program. *Human Genome Quarterly*, Spring, 1–2.

Beam, Alex. 1987. A Grand plan to map the gene code. *Business Week*, 27 April, 116–17.

Bell, George I. 1989. Workshop focuses on interface between computational science and nucleic acid sequencing. *Human Genome Quarterly*, Spring, 6, 8.

Berkowitz, Ari. 1989. Human genome program (letter). *Science*, 17 November, 874.

Blakeslee, Sandra. 1989. Discovery may help cystic fibrosis victims. *New York Times*, 24 August, B6.

———. 1989. Scientists develop new techniques to track down defects in genes. *New York Times*, 12 September, B6.

———. 1989. A road map for genes. *New York Times*, 10 October, C7.

Bower, B. 1989. Schizophrenia gene: A family link fades. *Science News*, 10 June, 359.

Burks, Christian, and John Lawton. 1988. Letter re: Release 1.0, LiMB database. 28 November, 2 pages.

Carrano, Anthony V. 1989. Physical mapping of human DNA: An overview of the DOE program. *Human Genome Quarterly*, Summer, 1–7.

Cherfas, Jeremy. 1982. *Man-made life*. New York: Pantheon Books.

Cooper, Dan M. 1989. Human genome program (letter). *Science*, 17 November, 874.

Crawford, Mark. 1989. Yeutter backs plan to map crop genes. *Science*, 3 March, 1137.

DeLisi, Charles. 1988. The human genome project. *American Scientist*, September, 488–493.

Dickson, David. 1988. Focus on the genome. *Science*, 6 May, 711.

———. 1988. A Soviet human genome program? *Science*, 8 April, 140.

———. 1989. Britain launches genome program. *Science*, 31 March, 1657.

———. 1989. European genome program delayed? *Science*, 24 March, 1548.

———. 1989. Genome project gets rough ride in Europe. *Science*, 3 February, 599.

———. 1989. UNESCO seeks role in genome projects. *Science*, 17 March, 1431–32.

———. 1989. Watson floats a plan to carve up the genome. 5 May, 521–22.

Diseases of the genome: An interview with Victor McKusick. *Journal of the American Medical Association*, 24 August 1984, 1041–48.

Drlica, Karl. 1984. *Understanding DNA and gene cloning: A guide for the curious*. New York: John Wiley & Sons.

Elmer-DeWitt, Philip. 1989. The perils of treading on heredity. *Time*, 20 March, 70–71.

Ezzell, Carol. 1989. The flatworm's turn. *Nature*, 29 June, 648.

Foote, Carole A. 1982. Designing better humans. *Science Digest*, October, 44.

Fox, Jeffrey L. 1987. Contemplating the human genome. *BioScience*, July, 457–60.

Frankel, Edward. 1979. *DNA: The ladder of life*. New York: McGraw Hill.

Gene machines breathe new life into the lab supply business. *Business Week*, 2 April 1984, 76.

Genetic finds may aid disease research. *Seattle Times*, 18 June 1989, A7.

Hall, Stephen S. 1988. Genesis: The sequel. *California*, July, 62–69.

———. 1990. James Watson and the search for biology's "holy grail." *Smithsonian*, February, 41–49.

Hoagland, Mahlon. 1981. *Discovery: The search for DNA's secrets*. Boston: Houghton Mifflin.

Hood, Leroy. 1988. Biotechnology and medicine of the future. *Journal of the American Medical Association*, 25 March, 1837–44.

The human genome organization (HUGO). 1988. Montreaux, Switzerland: HUGO.

Jaroff, Leon. 1989. The gene hunt. *Time*, 20 March, 62–67.

Johns Hopkins Medical Institutions. 1988. *Hopkins geneticist elected president of human genome organization*. Baltimore: Johns Hopkins Medical Institutions, Public Relations.

Kangilaski, Jaan. 1989. Looking to the future of genome mapping, sequencing. *Journal of the American Medical Association*, 21 July, 325.

Kanigel, Robert. 1987. The genome project. *New York Times Magazine*, 13 December, 44+.

Kolata, Gina Bari. 1980. The 1980 Nobel prize in chemistry. *Science*, 21 November, 887–89.

———. 1989. Scientists pinpoint genetic changes that predict cancer. *New York Times*, 16 May, B5, B12.

Koshland, Daniel E., Jr. 1989. Human genome program (letter). *Science*, 17 November, 873.

———. 1989. Sequences and consequences of the human genome. 13 October, 189.

Kristofferson, David. 1987. The BIONET electronic network. *Nature*, 5 February, 555–56.

Lange, Kenneth, and Michael Boehnke. 1982. How many polymorphic genes will it take to span the human genome? *American Journal of Human Genetics*, 842–45.

Leo, John. 1989. Genetic advances, ethical risks. *U.S. News and World Report*, 25 September, 59.

Lepkowski, Wil. 1988. Program to map entire human genome urged. *Chemical and Engineering News*, 15 February, 5.

Lewin, Roger. 1986. DNA databases are swamped. *Science*, 27 June, 1599.

———. 1986. Proposal to sequence the human genome stirs debate. *Science*, 27 June, 1598+.

———. 1986. Shifting sentiments over sequencing the human genome. *Science*, 8 August, 620–21.

———. 1987. National academy looks at human genome project, sees progress. *Science*, 13 February, 747–48.

———. 1988. Chance and repetition. *Science*, 29 April, 603.

———. 1988. Genome projects ready to go. *Science*, 29 April, 602–4.

———. 1989. Genome planners fear avalanche of red tape. *Science*, 30 June, 1543.

———. 1989. Mapping by color and x-rays. *Science*, 29 April, 425.

Lewis, Ricki. 1986. Computerizing gene analysis. *High Technology*, December, 46–50.

———. 1987. Digital imaging aids chromosome analysis. *High Technology*, January, 63.

Luria, S. E. 1989. Human genome program (letter). *Science,* 17 November, 874.

Mapping the human genome. *The Lancet,* 16 May 1987, 1121–22.

Marx, Jean L. 1985. Putting the human genome on the map. *Science,* 12 July, 150–51.

———. 1988. Evidence uncovered for a second Alzheimer's gene. *Science,* 16 September, 1432–33.

———. 1989. The cystic fibrosis gene is found. *Science,* 1 September, 923–35.

———. 1989. Detecting mutations in human genes. *Science,* 10 February, 737–38.

Mervis, Jeffry. L. 1989. BIONET bites the dust. *Science,* 14 July, 126.

Merz, Beverly. 1987. With current gene markers, presymptomatic diagnosis of heritable diseases is still a family affair. *Journal of the American Medical Association,* 4 September, 1132–33.

———. 1987. Mapping the human genome raises question: Which road to take? *Journal of the American Medical Association,* 4 September, 1131–32.

———. 1988. Gene mappers form collaborations. *Journal of the American Medical Association,* 4 November, 2477.

———. 1988. National research council endorses human gene mapping project. *Journal of the American Medical Association,* 11 March, 1433–34.

———. 1989. 700 genes mapped at world workshop. *Journal of the American Medical Association,* 14 July, 175.

Miller, Julie Ann. 1985. Lessons from Asilomar. *Science News,* 23 February, 122–23.

National Library of Medicine. 1988. *National biotechnology information center of the NLM.* Bethesda, Md.: National Institutes of Health.

Nelson, J. Robert. 1988. Genetics and theology: A complementarity? *The Christian Century,* 20 April, 388–89.

Olby, Robert. 1974. *The path to the double helix.* Seattle: University of Washington Press.

Palca, Joseph. 1989. Genome projects are growing like weeds. *Science,* 14 July, 131.

Paulson, Tom. 1989. Genetic treatment of illness closer in Seattle. *Seattle Post-Intelligencer,* 30 May, A1, A5.

Paulson, Tom. 1989. Mapping our makeup. *Seattle Post-Intelligencer,* 19 January, C1, C3.

Pines, Maya. 1978. *Inside the cell: The new frontier of medical science.* Bethesda, Md.: National Institute of General Medical Sciences.

———. 1984. *The new human genetics.* Bethesda, Md.: National Institute of General Medical Sciences.

———. 1987. *Mapping the human genome.* Bethesda, Md.: Howard Hughes Medical Institute.

Roberts, Leslie. 1987. Agencies vie over human genome project. *Science,* 31 July, 486–88.

———. 1987. Flap arises over genetic map. *Science,* 6 November, 750–52.

———. 1987. Human genome: Questions of cost. *Science,* 18 September, 1411–12.

——. 1988. Carving up the human genome. *Science*, 2 December, 1244–46.

——. 1988. Genome project. *Science*, 25 November, 1123.

——. 1988. NIH and DOE draft genome pact. *Science*, 23 September, 1596.

——. 1988. Race for the cystic fibrosis gene. *Science*, 8 April, 141–44.

——. 1988. Race for cystic fibrosis gene nears end. *Science*, 15 April, 282–85.

——. 1988. A sequencing reality check. *Science*, 2 December, 1245.

——. 1988. Watson may head genome office. *Science*, 13 May, 878–79.

——. 1988. Who owns the human genome? *Science*, 24 July, 358–61.

——. 1989. Ethical questions haunt new genetic technologies. *Science*, 3 March, 1134–36.

——. 1989. Genome mapping goal now in reach. *Science*, 28 April, 424–25.

——. 1989. Genome project under way, at last. *Science*, 13 January, 167–68.

——. 1989. New chip may speed genome analysis. *Science*, 12 May, 655–56.

——. 1989. New game plan for genome mapping. *Science*, 29 September, 1438–40.

——. 1989. Plan for genome centers sparks a controversy. *Science*, 13 October, 204–5.

——. 1989. Rifkin battles gene transfer experiment. *Science*, 10 February, 734.

Robertson, Miranda. 1986. The proper study of mankind. *Nature*, 3 July, 11.

Sandroff, Ronni. 1989. The baby shoppers. *Vogue*, May, 246–56.

Sanger, Fred. 1987. Reading the messages in the genes. *New Scientist*, 21 May, 60–62.

Schmeck, Harold M., Jr. 1988. DNA pioneer to tackle biggest gene project ever. *New York Times*, 4 October, 23, 26.

——. 1988. Evidence that schizophrenia is linked to genes. *New York Times*, 10 November, A3.

——. 1988. Gene studies emerging as key engine of science. *New York Times*, 6 September, C1, C6.

——. 1988. Hoping to repair defects, doctors take aim at gene targets. *New York Times*, 29 November, C3.

——. 1988. Momentum builds to map all genes. *New York Times*, 25 August, B28.

——. 1989. Group organizes to aid gene work. *New York Times*, 28 April, A4.

——. 1989. New methods fuel efforts to decode human genes. *New York Times*, 9 May, B5, B7.

——. 1989. Simplified gene-transplant method reported. *New York Times*, 2 June, A1, D14.

Schmidke, Jorg et al. 1986. Letter: Human gene cloning: The storm before the lull? *Nature*, 10 July, 319.

Scientists report finding gene for eye disease. *New York Times*, 18 July 1989, B7.

Sinsheimer, Robert L. 1983. Genetic engineering: Life as a plaything. *Technology Review*, April, 14–16.

Smith, Douglas. 1989. *Caltech chemists develop powerful, synthetic DNA-binding agents.* Pasadena: California Institute of Technology, Office of Public Relations.

Sohmeck, Harold M. 1987. New genetic map should aid battle against disease. *Seattle Post-Intelligencer*, 8 October, A3.

Spengler, Sylvia J. 1989. DOE human genome steering committee announced. *Human Genome Quarterly*, Spring, 3.

Stinson, Stephen. 1989. Debate swirls around human genome project. *Chemical and Engineering News*, 27 March, 6.

Sun, Marjorie. 1989. Consensus elusive on Japan's genome plans. *Science*, 31 March, 1656–57.

Swinbanks, David. 1989. Sequencing by committee. *Nature*, 29 June, 648.

Thompson, Larry. 1989. $100 million sought for study of genes. *Washington Post*, "Health Magazine." 10 January, 9.

Tivnan, Edward. 1988. Jeremy Rifkin just says no. *New York Times Magazine*, 16 October, 38–46.

Tressler, Arthur. 1988. *Cantor to direct LBL's human genome center.* Livermore, CA: Lawrence Livermore Laboratory, Press Release.

University of Washington Information Services. 1987. *Biochips may revolutionize information storage and transmission.* Seattle: University of Washington Information Services.

Weiss, Rick. 1988. Report adds to gene map momentum. *Science News*, 20 February, 117–18.

———. 1989. Predisposition and prejudice. *Science News*, 21 January, 40–42.

Wertz, Dorothy C., and John C. Fletcher. 1989. Disclosing genetic information: Who should know? *Technology Review*, July, 22–23.

Wickelgren, I. 1989. *MS gene discovery: A piece of the puzzle. Science News*, 8 July, 21.

Wingerson, Lois. 1988. Genes for sale. *Discover*, January, 85–86.

Woodward, Kenneth L. 1987. The genome initiative. *Newsweek*, 31 August, 58–60.

Wright, Susan, and Robert L. Sinsheimer. 1983. Recombinant DNA and biological warfare. *Bulletin of the Atomic Scientists*, November, 20–27.

Yarris, Lynn. 1987. *Human genome center for LBL.* Berkeley: Lawrence Berkeley Laboratory.

———. 1987. Science primer: The genetic process. *LBL Research Review*, Fall/Winter, 12–13.

———. 1987. Unlocking the genome. *LBL Research Review*, Fall/Winter, 2–11.

———. 1989. *First images of DNA helix obtained.* Berkeley, CA: Lawrence Livermore Laboratory, Public Information Department.

Zurer, Pamela. 1989. Panel plots strategy for human genome studies. *Chemical and Engineering News*, 9 January, 6.

Technical Articles and Reports

Botstein, David et al. 1980. Construction of a genetic linkage map in man using restriction fragment length polymor phisms. *American Journal of Human Genetics*, May, 314–31.

Bowcock, A. M. et al. 1988. Eight closely linked loci place the Wilson disease locus within 13q14–q21. *American Journal of Human Genetics*, November, 664–74.

Bowden, Donald W. et al. 1989. Identification and characterization of 23 RFLP loci by screening random cosmid genomic clones. *American Journal of Human Genetics*, May, 671–78.

Burke, David T. et al. 1987. Cloning of large segments of exogenous DNA into yeast by means of artificial chromosome vectors. *Science*. 15 May, 806–12.

Carrano, Anthony V. 1988. Establishing the order of chromosome-specific DNA fragments. In *Biotechnology and the Human Genome*, ed. Avril Woodhead and Benjamin J. Barnhart, 37–49. New York: Plenum Press.

Carrano, Anthony V. et al. 1989. A high-resolution, fluorescence-based, semi-automated method for DNA fingerprinting. *Genomics*, February, 129–136.

———. 1989. Constructing chromosome- and region-specific cosmid maps of the human genome. *Genome*, February, 1059–65.

Cavanaugh, Mark L. et al. 1988. *Human gene mapping library user's guide.* New Haven: Howard Hughes Medical Institute.

Chambon, Pierre. 1981. Split genes. *Scientific American*, May, 60–71.

Clark, Steven M. et al. 1988. A novel instrument for separating large DNA molecules with pulsed homogeneous electric fields. *Science*, 2 September, 1203–5.

Collins, Francis et al. 1987. Construction of a general human chromosome jumping library, with applications to cystic fibrosis. *Science*, 27 February, 1046–49.

Connell, C. et al. 1987. Automated DNA sequence analysis. *BioTechniques*, April, 342–48.

Cook-Deegan, Robert M. 1988. *Mapping our genes: Genome projects—how big, how fast?* Washington, D.C.: U.S. Congress Office of Technology Assessment.

DAP Computers. 1988. *Brochure: Active memory technology, inc.* Irvine, Ca.: Active Memory Technology, Inc.

DeLisi, Charles. 1986. *Memo: Information on human genome project.* Washington, D.C.: United States Department of Energy.

———. 1986. *Memo: Information on a major new initiative: Mapping and sequencing the human genome.* Washington, D.C.: United States Department of Energy.

———. 1988. Computation and the genome project: An historical perspective. *Proceedings: Interface between computational science and nucleic acid sequencing,* 12 December, 10–21.

Department of Energy Human Genome Steering Committee. 1988. *DOE human genome steering committee report, October 18, 1988.* Washington, D.C.: DOE Human Genome Steering Committee.

———. 1989. *DOE human genome steering committee report, January 16, 1989.* Washington, D.C.: DOE Human Genome Steering Committee.

Department of Energy, National Institutes of Health. 1988. *Memorandum of understanding.* Washington, DC: Department of Energy, National Institutes of Health.

Department of Energy Office of Energy Research. 1987. *Human genome initiative of the U.S. department of energy.* Washington, D.C.: DOE Office of Energy Research.

Department of Energy Office of Health and Environmental Research. 1986. *Genome Sequencing Workshop, March 3 and 4, 1986, Santa Fe, New Mexico.* Washington, DC: DOE Office of Health and Environmental Research.

Dracopoli, Nicholas C. et al. 1988. A genetic linkage map of 27 loci from PND to FY on the short arm of chromosome 1. *American Journal of Human Genetics*, October, 462–70.

Dulbecco, Renato. 1986. Turning point in cancer research: Sequencing the human genome. *Science*, 7 March, 1055–56.

Estvill, X. et al. 1989. Isolation of a new DNA marker in linkage disequilibrium with cystic fibrosis, situated between J3.11 (D7S8) and IRP. *American Journal of Human Genetics*, May, 704–10.

Gerber, Michael J. et al. 1988. Regional localization of chromosome 3-specific DNA fragments by using a hybrid cell detection mapping panel. *American Journal of Human Genetics*, October, 442–51.

Gilliam, T. Conrad et al. 1987. A DNA segment encoding two genes very tightly linked to Huntington's disease. *Science*, 13 November, 950–52.

Ginsberg, Michelle. 1988. *Database searching: A short comparative study.* London: Imperial Cancer Research Fund, Mutagenesis Laboratory.

Gough, Michael. 1986. *Technologies for detecting heritable mutations in human beings.* Washington, D.C.: U.S. Congress Office of Technology Assessment.

Hillis, W. Daniel. 1987. The connection machine. *Scientific American*, June, 108–15.

Howard Hughes Medical Institute. 1988. *Research in progress 1988.* Bethesda, Md.: Howard Hughes Medical Institute.

Human Gene Mapping Library, HHMI. 1988. *New Haven human gene mapping library chromosome plots #4, HGM9.5.* New Haven: Human Gene Mapping Library, HHMI.

Julier, Cecile, and Ray White. 1988. Detection of a NotI polymorphism with the pmet H probe by pulsed-field gel electrophoresis. *American Journal of Human Genetics*, January, 45–48.

Julier, Cecile et al. 1988. Linkage and physical map of chromosome 22, and some applications to gene mapping. *American Journal of Human Genetics*, February, 297–308.

Kornberg, Roger, and Aaron Klug. 1981. The nucleosome. *Scientific American*, February, 52–64.

Lake, James A. 1981. The ribosome. *Scientific American*, August, 84–97.

Lander, Eric et al. 1989. Study of protein sequence comparison metrics on the connection machine CM-2. *Proceedings: Supercomputing*, Vol II: Science and Applications, 2–12.

Lathrop, G. M. et al. 1984. Strategies for multilocus linkage analysis in humans. *Proceedings: National Academy of Sciences*, June, 3443–6.

Leppert, Mark et al. 1987. The gene for familial polyposis coli maps to the long arm of chromosome 5. *Science*, 4 December, 1411–13.

Litt, M. et al. 1988. Chromosomal localization of the human proenkephalin and prodynorphin genes. *American Journal of Human Genetics*, February, 327–34.

Little, Randall D. et al. 1989. Yeast artificial chromosomes with 200- to 800-kilobase inserts of human DNA containing HLA, Vk, 5S and Xq24–Xq28 sequences. *Proceedings: National Academy of Sciences*, March, 1598–1602.

McKusick, Victor A. 1988. *The morbid anatomy of the human genome*. Bethesda, Md.: Howard Hughes Medical Institute.

———. 1989. Mapping and sequencing the human genome. *New England Journal of Medicine*, 6 April, 910–14.

McKusick, Victor A., and Frank H. Ruddle. 1987. The status of the gene map of the human chromosomes. *Science*, 22 April, 390–405.

Nakamura, Yusuke et al. 1987. Variable number of tandem repeat (VNTR) markers for human gene mapping. *Science*, 27 March, 1616–22.

———. 1988. Localization of the genetic defect in familial adenomatous polyposis within a small region of chromosome 5. *American Journal of Human Genetics*, November, 638–44.

Olson, Maynard V. et al. 1986. Random-clone strategy for genomic restriction mapping in yeast. *Proceedings: National Academy of Sciences*, October, 7826–30.

———. 1989. A common language for physical mapping of the human genome. *Science*, 29 September, 1434–35.

Orita, Masato et al. 1989. Detection of polymorphisms of human DNA by gel electrophoresis as single-strand conformation polymorphisms. *Proceedings: National Academy of Sciences*, April, 2766–70.

Pittsburgh Supercomputing Center. 1989. *Biomedical computing initiative*. Pittsburgh: Pittsburgh Supercomputing Center.

Smith, Cassandra L. et al. 1987. A physical map of the escherichia coli K12 genome. *Science*, 12 June, 1448–53.

St. George-Hyslop, Peter et al. 1987. The genetic defect causing familial Alzheimer's disease maps on chromosome 21. *Science*, 20 February, 885–90.

Tasset, Diane M. et al. 1988. Isolation and analysis of DNA markers specific to human chromosome 15. *American Journal of Human Genetics*, June, 854–66.

Thinking Machines Corporation. 1988. *Connection Machine Applications*. Cambridge, Ma.: Thinking Machines Corporation.

——. 1988. *The connection machine family.* Cambridge, Ma.: Thinking Machines Corporation.

——. 1988. *Corporate Background.* Cambridge, Ma.: Thinking Machines Corporation.

——. 1988. *Technology Background.* Cambridge, Ma.: Thinking Machines Corporation.

Waltz, David L. 1987. Applications of the connection machine. *Computer*, January, 85–96.

White, Ray, and C. Thomas Caskey. 1988. The human as an experimental system in molecular genetics. *Science*, 10 June, 1483–88.

White, Ray, and Jean-Marc Lalouel. 1988. Chromosome mapping with DNA markers. *Scientific American*, February, 40–48.

——. 1988. Linked sites of genetic markers for human chromosomes. In *Annual Review of Genetics*, ed. Allan Campbell, 259–80. Palo Alto, CA: Annual Reviews, Inc.

White, Ray et al. 1985. Construction of linkage maps with DNA markers for human white chromosomes. *Nature*, 10 January, 101–5.

Yuzbasiyan-Gurkan, Vilma et al. 1988. Linkage of the Wilson disease gene to chromosome 13 in North-American pedigrees. *American Journal of Human Genetics*, June, 825–29.

Glossary

Adenine: One of the bases present as a nucleotide link in DNA or RNA. It pairs up with thymine in DNA and uracil in RNA.

Adenomatous polyp: Fleshy growths found in the colon.

Allele: An alternative form of a genetic locus. Alleles are inherited separately from each parent.

Amino acid: One of the 20 molecules which are the building blocks of proteins. The genetic code determines the sequence or arrangement of amino acids which make up a protein.

Aminoacyl-tRNA synthetase: Members of a class of enzymes which link particular amino acids with particular transfer RNA molecules. Each synthetase recognizes one specific type of transfer RNA and one specific type of animo acid.

Anticodon: A particular three-nucleotide region in transfer RNA which is complementary to a specific three-nucleotide codon in messenger RNA. The alignment of codons and anticodons is the basis for the arrangement of amino acids in a protein chain.

Arabidopsis thalania: A small weedy plant whose genome contains only about 70 million base pairs. Its genome is to be mapped and sequenced by researchers connected to the Genome Project.

Autoradiograph: A picture made with X rays to show radioactively labeled molecules or molecular fragments, and used to analyze the length and number of DNA fragments after they have been separated using gel electrophoresis.

Autosome: A chromosome not involved in sex determination.

Base: A flat ring structure made of nitrogen, carbon, oxygen, and hydrogen atoms, which functions as part of the building blocks of nucleic acids. The bases are adenine (A), cytosine (C), guanine (G), thymine (T), and uracil (U).

Base pair: Two bases, one in each strand of a double-stranded DNA molecule, which are attracted to each other by weak chemical interactions and link together.

Caenorhabditis elegans: A type of roundworm or nematode whose genome has been extensively studied. The total mapping of the genome of C. elegans is one of the goals of the Genome Project.

Cell: The smallest unit of living matter potentially capable of reproducing itself. A cell is bounded by a membrane which separates the inside of the cell from the outer environment. A cell contains DNA, where genetic information is stored; ribosomes, where proteins are made; and mechanisms for converting energy from one form to another.

Centimorgan: A unit of measure of recombination frequency. One centimorgan is equal to a one percent chance that a genetic locus will be separated from a genetic marker due to recombination in a single generation. In humans, one centimorgan is roughly equal to a physical distance of one million base pairs.

Chromosome: Subcellular structures in the nucleus of the cell which contain a long piece of DNA, plus the different proteins which organize and compact the DNA.

Chromosome jumping: A method of sequencing very large pieces of chromosomes which uses chromosomal fragments about 100,000 base pairs or more in length.

Chromosome walking: A method of sequencing a stretch of chromosome by laying out small fragments containing genetic markers so they overlap at the ends, and then moving ("walking") down the fragments from each genetic marker toward the center. Works for stretches of DNA less than 200,000 base pairs long.

Clone: 1. Noun: A group of identical cells, all derived from a single ancestor. Also, popularly used to refer to a particular member of such a group. 2. Verb: To undergo the process of creating a group of identical cells or identical DNA molecules derived from a single ancestor.

Codon: The three-letter word of the genetic code, made of three consecutive nucleotides whose order codes for one of the twenty amino acids. In certain cases several different codons code for the same amino acid. Several codons also act to stop or start signals for the creation of proteins.

Complementary base pairing: The process by which only certain nucleotides can align opposite each other in the two strands of DNA: A-T, A-U (in RNA), and G-C.

Contig: Groups of clones which represent overlapping (contiguous) regions of a genome.

Cosmid: A segment of DNA about 40,000 base pairs long.

Cosmid map: A form of physical genetic map using cosmids which gives the distance between genes and base pairs proportional to the number of bases. A cosmid map will pinpoint the location of genes to within ten thousand to a hundred thousand base pairs of each other.

Cytogenetic map: A type of physical genetic map which shows the banding patterns on chromosomes created with chemical stains, and which gives the rough location of genes on the chromosomes. Gross chromosomal abnormalities can be located to within ten million base pairs with a cytogenetic map.

Cytomegalovirus (CMV): One of a group of species-specific herpes virus. In humans, CMV lives in the salivary glands.

Cytosine (C): One of the bases which forms DNA or RNA and pairs with guanine.

Daughter molecule: A strand of DNA reproduced from an earlier DNA molecule.

Deoxyribonucleic acid (DNA): DNA; a long, thin chainlike molecule made of nucleotide units and outer "rails" of sugars and phosphates. The arrangement of the nucleotide units stores information necessary for life. DNA resembles a twisting double helix.

Diathesis: A predisposition to a certain disease, condition, or group of diseases.

Diploid: A full set of genetic material (two paired sets of chromosomes), one from each parental set. All but reproductive cells (that is, the autosomal cells) have a diploid set of chromosomes. In humans the diploid genome has 23 pairs of chromosomes, or 46 chromosomes in all. See Haploid.

DNA: See Deoxyribonucleic acid.

DNA sequence: The relative order of base pairs in a stretch of DNA, a gene, a chromosome, or an entire genome.

Dominant: In genetics, a trait or characteristic which will be expressed even though it is present in only one allele.

Drosophila melanogaster: The common fruit fly, an important experimental animal in genetics for nearly a century.

Eco RI: A restriction enzyme for E. coli discovered by Herbert Boyer and used in the first experiments in genetic engineering.

Electrophoresis: A method of separating large molecules, such as DNA fragments or proteins, from a mixture of similar molecules. An electric current is passed through a medium containing the mixture, and each kind of molecule travels through the medium at a different rate, depending on its electrical charge and size. Separation is based on these differences.

Enzyme: A protein molecule which accelerates certain biological chemical reactions.

Escherichia coli (E. coli): Important experimental bacteria in genetics, E. coli are commonly found in the digestive tracts of many mammals, including humans.

Eugenics: Attempts to improve hereditary qualities through selective breeding. Usually applied to humans.

Eukaryote: A major class of living beings, more highly organized than prokaryotes, possessing a well-defined nucleus containing DNA.

Evolution: A process of change in a certain direction. In science: the scientific theory, first presented in modern form by Charles Darwin and subsequently modified in certain details, that all life forms have their origin in preexisting types of organisms, and that the distinguishable differences are due to modifications in successive generations.

Exon: The protein-coding DNA sequences of a gene.

Familial polyposis coli: A medical syndrome or condition involving the presence of numerous polyps in a person's intestinal tract. A similar condition is called Gardner's Syndrome.

Familial adenomatous polyposis (FAP): A collective name for two genetically dominant conditions, familial polyposis coli and Gardner's syndrome, both characterized by the growth of hundreds of polyps in the intestinal tract. Usually appears by the time a person is 30 years of age. People with FAP have a high risk of developing colon cancer by age 40 if the polyps are not removed.

Gel electrophoresis: A particular method of separating DNA fragments or proteins. See Electrophoresis.

Gene: A small section of DNA which contains information for the construction of one protein molecule (or in some cases for making a molecule of transfer or ribosomal RNA).

Gene expression: The process of transferring information by means of messenger RNA from a specific region of DNA (a gene) to ribosomes, where a specific protein is made.

Gene therapy: The insertion of a normal gene directly into cells to correct a genetic defect.

Genetic cloning: The process of using microorganisms such as bacteria or viruses to produce millions of identical copies of a gene or genes.

Genetic code: The sequence of nucleotides, coded in triplets (codons) along mRNA, that determines the sequence of amino acids in the creation of a protein. The DNA sequence of a gene can be used to read the mRNA sequence, which in turn can be used to predict the amino acid sequence and thus the protein coded for by a gene. See Codon, Messenger RNA.

Genetic engineering: The manipulation of the information content of an organism to alter the characteristics of that organism. Genetic engineering may use simple methods such as selective breeding, or highly technological ones such as genetic cloning.

Genetic linkage: The proximity of two or more genetic markers on a chromosome; the closer the markers are, the less likely they will be separated during meiosis and thus the more likely they will be inherited together.

Genetic linkage map: A map of the relative positions of genetic loci on a chromosome, determined on the basis of how often the loci are inherited together. Distance is measured in centimorgans. A restriction map is a form of genetic linkage map.

Genetic locus: A position on a chromosome of a gene or other marker; also, the DNA at that particular position.

Genetic marker: An identifiable physical location on a chromosome, such as a restriction enzyme cutting site, a gene, or an RFLP marker, whose inheritance can be monitored. Genetic markers can be regions of DNA which express themselves in some known fashion by making a known protein. They can also be a segment of DNA with no known coding function but whose inheritance pattern can be determined.

Genetics: The study of the patterns of inheritance of specific traits.

Genome: All the genetic material in the chromosomes of a particular organism or species. The size of a genome is usually given in total number of base pairs. For example, the human genome is about 3 billion base pairs in length.

Genome projects: Research and technology development efforts aimed at mapping and sequencing some or all of the genome of human beings and other organisms. Collectively referred to as the "Genome Project."

Genomic library: A collection of clones made from a set of overlapping DNA fragments representing the entire genome of an organism.

Guanine: One of the bases which forms a part of DNA or RNA. Usually abbreviated with the letter "G," guanine pairs with cytosine.

Haploid: A single set of chromosomes (half the full set of genetic material) present in the egg and sperm cells of animals and in the pollen cells of plants. Humans have 23 chromosomes in their reproductive cells. See Diploid.

Heredity: The transmission of genetic characteristics from parents to offspring.

Heterozygous: Possessing different alleles at a genetic locus.

Homologous: Corresponding or similar in nature or position; describing regions of DNA which have the same nucleotide sequence, or genes with similar or identical alleles.

Homozygous: Produced by similar or nearly identical alleles.

Intron: A DNA sequence interrupting the protein-coding (exon) sequences of a gene, which is cut out of a sequence before it is translated into messenger RNA and coded into a protein. Sometimes referred to as "junk DNA."

Lac: An operon, or series of genes, found in E. coli, which were studied by Walter Gilbert and led to his development of a quicker way of sequencing DNA.

Lac repressor protein: The protein coded for by the lac operon in E. coli, which acts to turn off or repress the functioning of another gene.

Lambda: A particular phage (bacteria-attacking virus) used extensively in gene cloning.

Library: A collection of clones in no obvious order whose relationship can be established by physical mapping. See Genomic library.

Meiosis: The process of two cell divisions in the diploid predecessors of sex cells. Meiosis results in four daughter cells rather than the normal two, each of which has a haploid set of chromosomes. See Mitosis.

Messenger RNA (mRNA): RNA used to transmit information from a gene on DNA to a ribosome where the information is used to make a protein.

Met: A human oncogene found on chromosome 7.

Mitosis: A type of cell division in somatic cells in which the daughter cells contain the same number of chromosomes as the parent cell. Genes are equally distributed to all daughter cells, and a fixed number of chromosomes is maintained in all somatic cells of an organism.

Multilocus analysis: A method of genetic analysis which looks at the inheritance of several genetic loci at once.

Mutation: Any change in a DNA sequence that results in a new characteristic being inherited.

Nematode: Any of a class (*Nematoda*) of elogated cylindrical worms that are parasitic in animals or plants, or free-living in soil or water. One nematode, C. elegans, has been heavily studied by geneticists.

Nucleotide: One of the building blocks of DNA consisting of a base, a sugar molecule, and a phosphate molecule.

Oncogene: A gene with one or more forms associated with cancer. Many oncogenes are directly or indirectly involved with the regulation of cell growth.

Operon: A group of genes which occur together in DNA and are responsible for the control of some aspect of metabolism. Operons allow coordinated control of genes whose products have related functions.

Ordering: Placing a series of cosmids or other DNA fragments in proper linear order.

Parallel processing: A method of handling information in computers in which many different operations are performed at once by several computer chips, rather than one at a time by one chip.

Peptide: A small protein.

Phage: A virus that attacks bacteria. Sometimes called a bacteriophage.

Physical map: A map of the locations of identifiable landmarks on DNA, such as enzyme cutting sites, RFLP markers, or genes, regardless of inheritance. Distance is measured in base pairs.

Plasmid: A small circular DNA moleucle found inside bacterial cells which reproduces independently of the chromosomes. Plasmids reproduce every time the bacterial cells reproduces. Plasmids can transfer from one cell to another, even across species.

Polygenic disorder: A hereditary disorder resulting from the combined action of the alleles of two or more genes, such as heart disease, diabetes, and some cancers. Although inherited, such disorders depend on the simultaneous presence of several alleles; their hereditary patterns are more difficult to detect and to map. See Single-gene disorder.

Polymorphism: A difference in DNA sequence between individuals.

Prokaryote: A primitive cell or organism lacking a discrete nucleus or other subcellular structures. The DNA in prokaryotes is contained throughout the organism itself.

Protein: A class of long, chainlike molecules made of amino acids. Proteins usually are made of hundreds of amino acids; small proteins are usually called peptides. See Peptide.

Radiation hybrid mapping: A genetic mapping method combining procedures from genetic linkage and physical mapping methods. Genetic markers are separated by irradiating a chromosome with X rays. The resolution of the map produced depends on the amount of X ray dosage, with a theoretical upper limit to resolution of about 50,000 base pairs.

Recessive: In genetics, a trait or characteristic which will be expressed only if it is present in both alleles. A recessive gene will not express itself in the presence of its dominant allele.

Recombinant DNA: A DNA molecule containing strands from two or more distinct sources, formed using genetic cutting and splicing technologies.

Recombinant DNA technology: Procedures for joining together DNA segments in an environment outside of a cell or organism.

Recombination: The natural process of the breaking and rejoining of DNA strands to produce new combinations of DNA molecules and generate genetic diversity.

Restriction enzyme: Enzymes which cut DNA at specific nucleotide sequences.

Restriction fragment: Pieces of DNA created with restriction enzymes.

Restriction fragment length polymorphism (RFLP): A variation in the sizes of DNA fragments that have been cut by restriction enzymes. The polymorphic sequences responsible for RFLPs are used as markers on genetic linkage maps. See Genetic linkage map; Polymorphism; Restriction enzyme; Restriction map.

Restriction map: A form of genetic linkage map which uses RFLPs as markers. Restriction maps can have a resolution of about a million base pairs.

Retinitis pigmentosa (RP): A genetically caused disease characterized by the degeneration of cells in the retina, progressing from night blindness to tunnel vision to complete blindness. Usually first appears in childhood. A genetic marker for one form of RP has been found on chromosome 3.

Ribonucleic acid (RNA): A long, thin chainlike molecule usually found as a single chain, made of the nucleotides adenine (A), cytosine (C), guanine (G) and uracil (U), and playing a vital role in the transformation of genetic information from DNA into proteins. There are several types of RNA, including messenger RNA, ribosomal RNA, and transfer RNA.

Ribosomal RNA (rRNA): A class of RNA found in the ribosomes of cells.

Ribosome: A large ball-like structure in a cell which is made of rRNA and about 50 specific ribosomal proteins and acts as the "workbench" for the creation of proteins.

RNA: See Ribonucleic acid.

RNA polymerase: The enzyme complex responsible for making RNA from DNA.

Saturation mapping: A method of comparing random pieces of DNA from a particular chromosome with DNA from individuals afflicted with a specific genetic disease, in order to find matches between the two and determine the chromosomal location of the suspect gene or genes.

Sequence map: A physical map of the actual base-pair sequence of an entire genome, the complete spelling of all the genetic instructions in a genome, called by some "the ultimate [genetic] map."

Sex chromosomes: The X and Y chromosomes in human beings that determine the gender of an individual. Human females have two X chromosomes (XX) in their diploid cells. Human males have an X and Y chromosome (XY) in their diploid cells.

Simian virus 40 (SV40): a virus which infects the cells of monkeys and can transform them into cancerous tumors, used in the first genetic engineering experiments by Paul Berg.

Single-gene disorder: A hereditary disorder caused by a single gene, such as Duchenne muscular dystrophy, retinoblastoma, or sickle-cell anemia. See Polygenic disorder.

Somatic cell: Any cell in an organism which is not a sex or reproductive cell. Somatic cells reproduce by mitosis.

Somatic cell hybridization (SCH): A technique for cloning human chromosomes in which the tumor cells of humans are fused with those of another species, such as mice, using certain chemicals, an electric field, and/or the Sendai virus. The hybrid cell lines eventually end up containing eight to twelve of the 46 human chromosomes, in addition to the remaining rodent chromosomes. Researchers then take a large set of somatic cell hybrids containing different combinations of chromosomes and compare the presence or absence of a specific human chromosome with the presence or absence of a particular gene.

Thymine (T): one of the bases which forms a part of DNA. Pairs with adenine (A) in DNA. Thymine is not found in RNA; uracil is the base pair instead.

Transcription: The synthesis of mRNA from a gene (a sequence of DNA); the first step in gene expression.

Transfer RNA (tRNA): a class of small RNA molecules having structures with triplet nucleotide sequences that are complementary to the codons in mRNA, and which position the amino acids in the correct order during protein synthesis.

Translation: The process of converting the information in mRNA into protein. Translation follows transcription.

Uracil (U): one of the four bases that forms RNA; base pairs with adenine (A). Not usually found in DNA, which uses thymine instead.

Virus: A class of infectious agents usually composed of DNA or RNA surrounded by a protective protein coat. Much smaller than cells and unable to reproduce on their own, viruses must commandeer the reproductive machinery of other cells in order to reproduce.

YAC: See Yeast Artificial Chromosomes.

Yeast: One of many species of a one-celled organism containing a true nucleus, with biochemical properties very similar to those of higher organisms; used extensively in genetic engineering.

Yeast Artificial Chromosomes (YAC): very large pieces of DNA from another species spliced into the DNA of a form of yeast, and used in mapping the genome of different species. YACs are large, from 200,000 to 600,000 base pairs long.

Index